# ION CYCLOTRON
# RESONANCE
# SPECTROMETRY

# ION CYCLOTRON RESONANCE SPECTROMETRY

THOMAS A. LEHMAN
Associate Professor of Chemistry
Bethel College
North Newton, Kansas

MAURICE M. BURSEY
Professor of Chemistry
University of North Carolina at Chapel Hill

A WILEY-INTERSCIENCE PUBLICATION

JOHN WILEY & SONS, New York · London · Sydney · Toronto

*Library of Congress Cataloging in Publication Data:*

Lehman, Thomas A.      1939–
    Ion cyclotron resonance spectrometry.

    "A Wiley-Interscience publication."
    Bibliography: p.
    Includes indexes.
    1.  Ion cyclotron resonance spectrometry.   I.   Bursey, Maurice M., joint author.   II.   Title.

QD96.I54L44        541′.39′028        76–26533
ISBN 0–471–12530–X

Printed in the United States of America

10 9 8 7 6 5 4 3 2 1

To DICK and JOAN

# PREFACE

The development of ion cyclotron resonance (ICR) spectrometry has been so extensive that it is now seen in most of the prominent journals of chemistry and chemical physics. Its varied consequences are important enough that the broad exposition in this book should draw the attention of chemists from many quarters.

The foundations of ICR spectrometry rest on electromagnetic theory. Not all who wish to study ICR or to understand its results can approach the subject from this direction. To meet their needs, Chapter 1 presents the technique in terms of the simple physics that all chemists have studied, except for a brief excursion into Fourier transform ICR spectrometry.

Readers without prior knowledge of ICR will benefit from the clear indications of synonymous terms, the careful use of electromagnetic units, the frequent references to mass spectrometry, and the comparisons to solution chemistry.

Reaction chemistry is developed in Chapters 2 and 3, first in generalizations, then in a systematic exposition by reaction types. Much attention is given to a comparison of reactions in solution and in the gas phase. Some remarkable insights into solvent effects have emerged, and many more are expected.

Highlights of the final chapters include a very careful presentation of the Langevin and Gioumousis-Stevenson equations, a variety of ways to calculate rate constants, a review of the effects of translational and internal energy on reaction rates, and a look at what happens when light interacts with the reacting ions and molecules.

Proper attention is given to some of the early methods that are being supplanted by newer experiments, although the approach is not historical. Examples include magnetic field modulation, rate-constant measurements in the drift cell, and proton affinity measured in the drift cell. New researchers will want to understand how such older results were obtained even if similar data are now more easily produced by other methods.

Research workers will find the Bibliography very useful; in it we give the full literature reference to every publication through 1973 that could be found on ICR spectrometry. The list was compiled with the aid of many research directors and a computer search of publication titles. To this we have added some more references from 1974 and 1975 without attempting to be exhaustive.

Arrangement of publications in the Bibliography is by type of experiment, with each paper assigned to one of six categories. This is convenient but becomes somewhat arbitrary for papers that report findings in more than one category. If a paper reports rate constants, it will very likely be found under "Analysis of Rates," no matter what else it contains. The utility of the Bibliography is further enhanced by an author index.

Two numbering schemes are used for footnotes. A letter followed by a number (e.g., A1) refers to the Bibliography, while a number (e.g., 1) refers to the end of the chapter. This has allowed us to eliminate repetitive entries and cross references by maximizing our use of the Bibliography.

The book stays close to the interests of a hypothetical "general chemist" throughout. Ion cyclotron resonance, as developed by Graham, Malone, and Wobschall, is excluded because it has not been the method of choice for the study of chemical reactions. Also unexamined are topics essential to the design of the ICR spectrometer, such as ion optics and the complicated circuitry of the detection system. An appendix deals with one aspect of the latter; otherwise we have treated the electronic components as black boxes. Although ion-molecule reactions are our chief interest, we look at them mainly as chemistry, and leave almost entirely to others an exposition of this fascinating area of research in the language of chemical physics. Finally, the book is neither a literature review nor a manual on the techniques of ICR spectrometry.

We began work on the book in 1970 when we were together in the laboratory at Chapel Hill. Since that time one of us has moved two hundred meters into a new building, while the other continued his writing in Belgium, Zaire, and Kansas. Happily, our collaboration has survived these major relocations, although they have cost some time.

The list of those who have influenced the contents of the book has grown past the point of individual recognition. We thank them all. Two who read the entire manuscript and provided pages of detailed criticism were J. L. Beauchamp and J. H. Futrell. We and readers of the book owe them much. We also thank the University of North Carolina and Bethel College for support of some of the manuscript-production expenses.

THOMAS A. LEHMAN
MAURICE M. BURSEY

*North Newton, Kansas*
*Chapel Hill, North Carolina*
*May 1976*

# CONTENTS

# ION CYCLOTRON
# RESONANCE
# SPECTROMETRY

## CHAPTER 1

Nature of the
Technique

## INTRODUCTION

Reactions of ions with molecules in solution constitute much of
classical and modern chemistry.  Our understanding of such
reactions has evolved from the early observations of stoichiometry to Faraday's quantitative studies of ions in solution and
the current required to produce or discharge them, and on to the
inferences about molecular shapes and the theories of acids and
bases.  The great impact of modern instruments on solution
chemistry is well illustrated by noting that a chemist today
seeks to discuss an overall reaction in terms of the approximate
geometry of the transition states of each elementary reaction, a
certain knowledge of the atom(s) transferred, and perhaps information about the arrangement of the solvent molecules around the
ions before and after the reaction.
     Solvent molecules can greatly influence the nature and
extent of a reaction, changing the rate by orders of magnitude,
and sometimes changing the products.  Among the physical properties of a solvent, the polarity of its molecules is of particular importance.  An ionizing solute will find it much easier to
ionize in a polar solvent, whose molecules orient themselves
around the ions and thereby reduce the electrostatic potential
energy of the system.  In the case of salts, the free energy of
the dissolved cations and/or anions must be reduced by electrostatic interaction with solvent molecules, or the salt will
precipitate.
     It is well known that water will dissolve large amounts of
some gases; this requires an extensive interaction between
solute and solvent.  An example is shown in Figure 1-1, in which
the enthalpy of the hydrogen-chlorine system is shown in several
states.  It should be noted that enthalpy is highest in the
gaseous ions and lowest in the ions in solution.  Here they are
separated by more than 15 eV.  The large differences between the
energies of gaseous and dissolved ions lead us to expect that

1

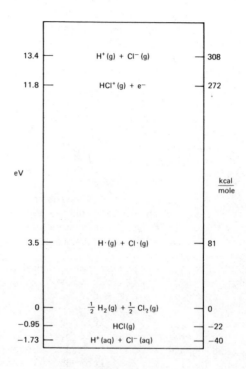

Figure 1-1. Enthalpy diagram for the formation of $H^+$, $Cl^-$, and $HCl^+$ ions. The enthalpy of the aqueous ions is at infinite dilution.

their reactivities will be quite different. The ions that react with molecules in solution are always under the influence of the solvent to some extent.[1]

Reactions between ions and neutral molecules in the gas phase have been actively studied for about 25 years[2,3]; the results make interesting comparisons and contrasts with solution reactions. The mass spectrometer, in which ions are produced by electron impact, is the logical means of generating, handling, and analyzing gaseous ions, usually cations. Vacuum technology has developed to the point that mass spectrometers are normally

operated at pressures no greater than a μtorr; collisions between ions and molecules occur to a negligible extent.  The mass
spectrum obtained under such conditions generally is uncluttered
by secondary ions, namely ions formed by reactive ion-molecule
collisions.  It was not always so.  J. J. Thomson observed a
signal corresponding to m/e = 3 and correctly assumed it to be
$H_3^+$.[4]  However such secondary ions came to be regarded as a
nuisance to be eliminated by the construction of better vacuum
pumps, and did not receive the serious attention of chemists
until many years later.

Mass spectrometers have been built to operate at somewhat
higher pressures in the ion source, so that reactive collisions
occur.  The low pressure necessary for mass analysis is maintained in the analyzer by attaching to it a very large pumping
system.  The source has its own pump; in one such system[5] the
pressure in the source can be $10^3$-$10^4$ times as great as in the
analyzer.  The narrow slit through which ions pass from source
to analyzer is the only connection between the two regions.
Source and analyzer are said to be *differentially pumped*.

THE TANDEM MASS SPECTROMETER

Two mass spectrometers can be joined to study bimolecular ion-
molecule reactions.  Reactant ions can be selected by mass and
their kinetic energies varied at will.  In the first spectrometer, ions are formed and separated, and those of the desired
mass pass through a narrow slit into a space containing
molecules at a pressure sufficiently high to allow most ions to
collide with one or more molecules.  The product ions, which
result from reactive collisions, are swept out and analyzed by
the second instrument.

A tandem mass spectrometer is shown in Figure 1-2.  In this
instrument the ion source is a double-focusing mass spectrometer.  As is always true in mass spectrometry, the ions are
accelerated as the first step in mass analysis.  A deceleration
lens reduces the kinetic energy of the reactant ions and sends
them into the collision chamber with an energy that is variable
from 0.2 to 100 eV, with an energy spread of 0.3 eV.  Here many
of them undergo reactive collisions.  Ions are drawn out of the
collision chamber by an accelerating lens.  They enter a quad-
rupole mass spectrometer, which changes the shape of the beam
prior to the final analysis.  This is accomplished by a second
double-focusing mass spectrometer.[6]

Tandem mass spectrometers have been built in various
laboratories, and have provided some of the most detailed information available on the chemistry of ion-molecule reactions.[7]

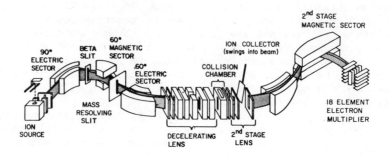

Figure 1-2.  Schematic of a tandem mass spectrometer
built at the Aerospace Research Laboratories.  Courtesy
of Dr. T. O. Tiernan.

Most types of reaction mentioned in this book were identified by
workers using high-pressure or tandem mass spectrometers.  Many
chemists have not had access to a tandem instrument because none
has ever been put into commercial production.

ION CYCLOTRON RESONANCE SPECTROMETRY

Ion cyclotron resonance spectrometry[A4,E1,E6,E9,A12,A35] is a
technique for studying gaseous ion-molecule reactions.  The
first ICR spectrometer was built by Varian Associates and
delivered to Stanford University in 1966, and a commercial
instrument soon became available.[8]  In this technique the ions
are also produced by electron impact, but their subsequent paths
and means of analysis differ greatly from those of the mass
spectrometer.

The idea that the cyclotron principle could be used with
certain modifications so as to obtain high mass resolution is
due to Sommer, Thomas, and Hipple.  Their instrument, which they
called an omegatron and used to measure m/e for the proton, was
built around 1950 at the National Bureau of Standards.[9]

The ICR cell is shown in Figure 1-3.  It is placed between
the poles of an electromagnet of flux density B, directed as
shown.[10]  Electrons are emitted from a hot rhenium filament, are
accelerated toward the near side of the cell by imposing a
negative potential on the filament, are collimated by B, and
pass through the cell.  A small fraction of them undergoes

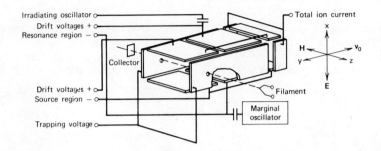

Figure 1-3. Cutaway view of ICR cell. (Reprinted, by permission, from *Ann. Rev. Phys. Chem. 22*, 527 (1971), Figure 1.)

ionizing collisions with molecules: $M + e^- \rightarrow M^+ + e^- + e^-$. The electron current falling on the collector plate is measured as the emission current. The potential difference between the filament and the side of the cell determines the ionizing energy.

The side plates of the cell carry a small positive potential ($\lesssim 0.5$ V) that keeps the ions from drifting to the sides of the cell, there being no restraint on such motion due to the magnetic field. It is called a trapping potential.

DYNAMICS OF NONRESONANT IONS

The cyclotron equation can be applied to the ions. For an ion of mass m, acceleration a, charge[11] e, and velocity v perpendicular to B, the force is

$$F = ma = evB$$

and will be perpendicular to both B and v.[12] Circular motion results for sufficiently large values of B; for this motion, $a = v^2/r$, where r is the radius of the ion path. Thus

$$ma = mv^2/r = evB$$

If $\omega_c$ is the angular frequency of the ion in rad/s, then

$$mv/r = m\omega_c = eB$$

or

$$\omega_c = eB/m \tag{1}$$

This is the basic cyclotron equation, and contains the important result that the angular frequency $\omega_c$ is independent of v.

When the detection of ions of a given m/e is considered (page 8), the frequency $\nu_c = \omega_c/2\pi$ will be more important than the angular frequency.[13]

$$\nu_c = eB/2\pi m$$

or

$$m/e = B/2\pi\nu_c \tag{2}$$

The second form of this equation shows directly that at constant frequency, m/e varies linearly with B. A linear mass scale for resonant ions is achieved by varying B, the magnetic flux density.

The frequency of $CO_2^+$ in a magnetic flux density of 1.0 T[14] is

$$\nu_c = \frac{1.6 \times 10^{-19} \text{ C} \times 1.0 \text{ T}}{6.28 \times 44 \text{ u} \times 1.66 \times 10^{-27} \text{kg/u}} = 3.5 \times 10^5 \text{ Hz} = 350 \text{ kHz}$$

that is, a $CO_2^+$ ion completes $3.5 \times 10^5$ circular orbits per second in a flux density of 1 T.

The ions are made to drift from the source to the resonance region[15] of the cell by the presence of a potential difference between plates above and below the electron beam. The electric field intensity (a vector) between the plates is perpendicular to the magnetic field vector, and these crossed fields cause the ions to move at a right angle to both fields. The drift velocity is given by

$$v_{drift} = \varepsilon_{drift}/B \tag{3}$$

where $\varepsilon_{drift}$ is the electric field intensity, the quotient of the potential difference by the distance between plates. The relationship between the three vectors involved is shown in Figure 1-4.

The distance between upper and lower plates is about 1 cm in the most widely used cell, and the potential difference between plates is in the neighborhood of 0.5 V. Thus in a field of 1.0 T,

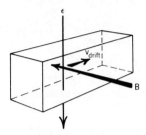

Figure 1-4.  Vector contributions to the drift motion.
For a derivation of the equation v = (ε/B) and its
application as a velocity selector, see the work by
Weidner et al.[16]

$$v_{drift} = \frac{0.5 \text{ V}}{0.01 \text{ m} \times 1.0 \text{ T}} = 50 \text{ m/s}$$

From the electron beam to the far end of the analyzer region is
a linear distance of about 8.6 cm, so the ion spends 0.086/50,
or about $2 \times 10^{-3}$ s in the cell under these conditions.

We can now gain an idea of the length of the cyclotron path
of a specific ion as it moves through the cell.  This length is
the product of ion speed and drift time.  It is argued below
that a molecular ion has nearly the same translational energy as
the molecule from which it was formed, unless the ion is absorbing energy from a radiofrequency oscillator.  For $CO_2$ at room
temperature, this energy corresponds to a speed of 410 m/s.  The
distance traveled in $2 \times 10^{-3}$ s by a $CO_2^+$ ion at this speed is
0.8 m.

The result, $2 \times 10^{-3}$ s, for the time an ion resides in the
cell has only order-of-magnitude significance, for several
reasons.  It depends directly on B, which in a scan of typical
m/e values would be swept from about 3000 to 13 000 gauss.  Thus
the drift time through the cell varies considerably.  An ion
that reacts to form a heavy product ion will have a relatively
long time to do so, because, by Equation (2), the cyclotron
frequency of the heavy ion will occur at a high value of the
magnetic field.  In contrast, ion flight times in a mass spectrometer are roughly a few μs; this also depends on instrumental
parameters and the mass of the ion.

Two other simple calculations can be done to describe the

trajectories of nonresonant ions. For $CO_2^+$ at room temperature
in a magnetic flux density of 1.0 T, for which the frequency of
350 kHz was obtained above, the diameter of the ion path is

$$D = \frac{v}{\pi \nu} = \frac{410 \text{ m/s}}{3.14 \times 3.5 \times 10^{-5}/s} = 0.4 \text{ mm}$$

The thermal velocity v, and hence the diameter D, decrease as
the ionic mass increases.

During one revolution, the distance this same ion drifts
through the cell is

$$d = \frac{50 \text{ m/s}}{3.5 \times 10^5/s} = 0.014 \text{ mm}$$

Because D is much larger than d, the orbits of the ions are very
nearly circular.


CYCLOTRON RESONANCE DETECTION

An electrical oscillator is a circuit in which the voltage
varies sinusoidally about a central value. The frequency of the
oscillation depends on the inductance of the coil and the size
of the capacitor in the circuit. Such circuits store energy as
they oscillate. The detection of ions in ICR spectrometry is
done by incorporating the upper and lower plates of the resonance region into the capacitance of a marginal oscillator.[17]
Ions are detected when their frequency, as defined by Equation
(2), matches (is in resonance with) that of the oscillator
circuit. They absorb energy from the radiofrequency field
between these plates, which creates an electrical signal in the
circuit. A marginal oscillator is used because of its great
sensitivity to such signals. An ion in resonance absorbs
energy, gains speed, and moves through a longer path, as shown
in Figure 1-5.[18] Equation (1) shows that $\omega_c$ (and hence $\nu_c$) is
independent of v and r separately but depends only on their
ratio. An ion remains in resonance by increasing its radius in
proportion to its increase in velocity.

It is convenient to choose a frequency such that a change
of 100 gauss (0.01 T) in B corresponds to a change of one atomic
mass unit (u). Using Equation (2),

$$\nu_c = \frac{1.6021 \times 10^{-19} \text{ C} \times 0.01 \text{ T}}{6.2832 \times 1 \text{ u} \times 1.6604 \times 10^{-27} \text{kg/u}}$$

$$= 1.5357 \times 10^5 \text{ Hz} = 153.57 \text{ kHz}$$

Figure 1-5.  Motion of a resonant ion in a trapping
cell.


The signal/noise ratio is greatly enhanced by signal modulation and phase-sensitive detection.[19]  Either some resonance
condition or the supply of ions is interrupted at a given frequency, 27 Hz in the commercial instrument, and a phase-
sensitive detector compares the signals so obtained.  Among the
means for changing the signals are:  (a) the trapping voltage
$V_T$ is modulated by a square wave on one trapping plate, (b) the
energy of the ionizing electrons is modulated, usually from
below to above the ionization potential,[A3] (c) the magnetic
field is modulated over a range of a few gauss, producing a
derivative scan similar to electron spin resonance spectrometry,[20] (d) a grid is placed between the filament and the cell
and a voltage is pulsed on it to keep electrons from reaching
the cell,[A8] and (e) a second radiofrequency oscillator is
switched on and off for double resonance (see below).  All are
illustrated in Figure 1-6.  Of these, modulation of the trapping
potential is the most common.  Field modulation is done by
Helmholtz coils placed on the magnet pole caps.  It was standard
on the first commercial instruments, but is less sensitive than
some other methods.

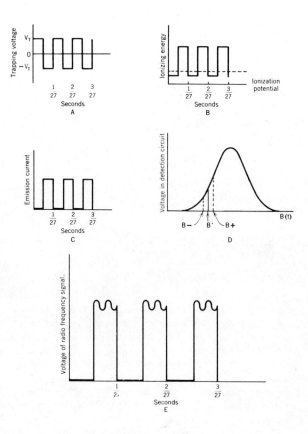

Figure 1-6.   Ion pulsing and signal-modulation schemes.
(a) The trapping voltage is pulsed. When it is at $-V_T$,
ions leave the cell.   (b) The ionizing energy is pulsed.
At the low value no ions are formed.   (c) B(t) sweeps
across the absorption curve.  Helmholtz coils, set just
inside the magnetic pole caps, repeatedly impose a
small incremental field B+, then an equal decremental
field B- on B(t).   The detector compares the signal at
B+ to that at B-.   As B sweeps through the curve, the

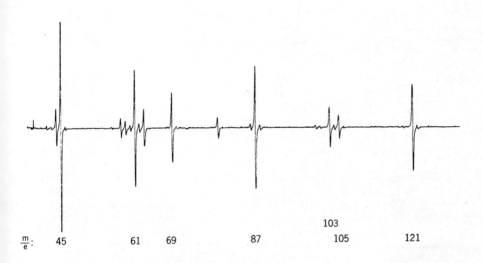

$\frac{m}{e}$:    45            61    69            87          105        121

Figure 1-7.  Scan of 2-propanol from m/e 35 to 135.
The sample pressure is 20 μtorr (uncalibrated ion-gauge
measurement) and the energy of the ionizing electrons
is 12.4 eV.  Every signal for which m/e >60 is the
product of an ion-molecule reaction.


    In theory, the detector can indicate the presence of as few
as 14 ions,[A35] which for an analyzer region 6 cm long and a
drift velocity of 50 m/s corresponds to a current of 2 x $10^{-15}$
A.
    An ICR scan of 2-propanol is shown in Figure 1-7.  It was
obtained at a pressure of about 20 μtorr and an ionizing energy
of 12.4 eV with modulation of the magnetic field.  The signal at
the molecular mass 60 is very small.  Every signal at values of

---

    detector produces a derivative trace.  To see this,
    note that the difference between B+ and B- is most
    positive when B is at the point of greatest slope on
    the low-field side of the absorption curve, most nega-
    tive when B is at the corresponding point on the high-
    field side, and zero when B is at the highest point in
    the curve.  (d) The number of ionizing electrons pass-
    ing through the cell is pulsed.  (e) A second radio-
    frequency oscillator is pulsed; it energizes possible
    reactant ions (see text).

m/e greater than 60 indicates the product of an ion-molecule
reaction.   The ion at m/e 45 is $C_2H_5O^+$; it is well known in conventional mass spectrometry.   The ion at m/e 121 is $(C_3H_7OH)_2H^+$,
the protonated dimer of the neutral molecule.


ION CURRENT AND PRESSURE MEASUREMENT

The ion current is measured in the region of the cell farthest
from the source.   The side plates in this region are at ground
potential, and in the absence of a trapping potential the ions
move outward to them.   The top and bottom plates are also sometimes at ground, but better control over the ions is maintained
if these plates carry a drift potential.   This is best done by
wiring them to the respective source plates, as shown in Figure
1-3.   This avoids the pickup of a radiofrequency signal from the
analyzer drift plates, which would be undesirable.   The current
is measured by an electrometer, and is a few pA at most.

     In an ICR spectrometer the ion collection is done after the
ions have been mass-analyzed.   In a mass spectrometer the
measurement of total ion current is done before the ions are
separated, and this intercepts the separation process.   The
measurement of the number of ions of a particular m/e is done by
collecting them.


*Pressure Measurement*

The neutral gas is admitted continuously to the region of the
cell via an adjustable leak valve.   An ion pump removes molecules by ionizing them and attracting the ions to a cathode.
The cathode current is proportional to the pressure, but the
constant of proportionality (ionization efficiency) depends on
the gas.

     The pressure of the neutral gas does not need to be known
accurately in ICR studies of reaction chemistry.   For this
purpose the pressures measured as ion pump currents are satisfactory.

     Accurate pressures are required for the measurement of rate
constants.   For this a capacitance manometer[21] is convenient.[C5]
The gas whose pressure is to be measured is in contact with a
metal diaphragm under tension.   Electrodes carrying an AC signal
are placed on either side of the diaphragm.   The diaphragm and
electrodes constitute a capacitor.   Changes in pressure cause a
change in the position of the diaphragm and a corresponding
change in capacitance.   This results in a change in the amplitude of the AC signal.

When the low-pressure limit ($\leq 10^{-5}$ torr) must be extended, the capacitance manometer can be used to calibrate an ion gauge[C61] for each gas.

## ION CYCLOTRON DOUBLE RESONANCE

A distinguishing feature of ICR spectrometry is the connection that can be made between a product ion and the reactant ion that formed it in an ion-molecule collision. This is done by using a variable radiofrequency oscillator to add energy[22] to possible reactant ions while the product ion is in resonance with the detecting oscillator. It is apparent from Equation (2) that $m\nu_c$ is a constant when an ion is in resonance at the required value of B. Ions of a different mass m' can be brought into resonance at that value of B by supplying the frequency $\nu'$ such that $m\nu_c = m'\nu_c'$. If m' is reacting to form m, heating m' will cause a change in the amount of m, and this is observed by the detecting oscillator. The technique is known as ion cyclotron double-resonance (ICDR) spectrometry.[A1,E1,E9]

There are several reasons for the double-resonance phenomenon, that is, the change in signal intensity of the product when the reactant ion is irradiated. A reaction rate constant is a function of temperature. Because ions and molecules do not normally reach thermal equilibrium before reaction in an ICR cell, it is more appropriate to speak of the dependence of the rate constant on ion velocity or energy, usually denoted $dk/dE$.[23]

Double-resonance signals can also arise for instrumental rather than chemical reasons. Ions can absorb enough energy from the variable frequency oscillator so as to come in contact with the upper or lower plates of the cell, as suggested by Figure 1-5. Ions thus removed are said to be swept out of the cell.[C21] A double-resonance signal is sometimes accompanied by a change in total ion current; if ion current is measured as a function of frequency during a double-resonance scan, the current changes at those frequencies for which double-resonance signals are seen. For example, a decrease in the level of the product ion signal may be due at least in part to the partial sweep-out of reactant ions. If the sign of $dk/dE$ is positive, these two contributions to the double-resonance signal will offset each other in part. In order to study $dk/dE$, double-resonance signals should be examined at the lowest possible radiofrequency fields after they have been located at higher fields.[E6,E26]

The sweep-out effect can give important chemical information. The link between reactant and product ion appears as a

decrease in the number of product ions when the potential reactant ions are swept to the edge of the cell. The difficulty is
the tendency to *assume* that a negative ICDR signal represents a
negative dk/dE. The sweep-out effect has been systematically
exploited, as will be shown in Chapter 5.

The double-resonance oscillator[24] can send energy into
either the source or the resonance region. Double-resonance
signal intensities will generally be different in the two
regions. Suppose that a double-resonance signal obtained in the
resonance region indicates that the reactant is a primary ion.
(A primary ion is one formed in the initial ionization process--
either the molecule-ion or a lighter ion formed from it by fast
unimolecular decomposition.) If the same reaction is studied in
the source, the signal may well be more intense, because there
the primary ions have not been greatly depleted by reaction.
Conversely, if a double-resonance signal in the resonance region
identifies as the reactant an ion that is produced by an ion-
molecule reaction, that double-resonance signal will almost
certainly be much weaker in the source because the reactant ion
will not be present there in sufficient numbers.

Figure 1-8 displays an ICDR experiment on an ion of mass
141 produced from DMSO. The marginal oscillator frequency is
103.23 kHz and the double-resonance signal appears at 187 kHz.
The mass of the reactant ion is thus $m' = m\nu_c/\nu_c' = 141$ x
(103.23/187) = 78. This is the mass of the DMSO molecular ion.
It can be assumed that this ion reacts with a neutral DMSO
molecule, mass 78. The reaction is thus

$$CH_3\text{-}\overset{\overset{O}{\|}}{S}\text{-}CH_3^{+\cdot} \;+\; CH_3\overset{\overset{O}{\|}}{S}\text{-}CH_3 \;\longrightarrow\; CH_3\text{-}(\overset{\overset{O}{\|}}{S}\text{-}CH_3)_2^{+} \;+\; \cdot CH_3$$

$$\quad 78 \qquad\qquad\quad 78 \qquad\qquad\qquad 141 \qquad\qquad 15$$

Mass balance requires a neutral product of mass 15.


DYNAMICS OF RESONANT IONS

The interaction of ions with an alternating electric field is
fundamental to ICR spectrometry, in both the detection of ions
and their irradiation in double-resonance experiments. Because
the detection system depends on the power drawn by the ions from
the radiofrequency signal sent into the resonance region, the
equation for power absorption that we are about to derive is of
great interest.

The radiofrequency (designated by subscript rf in equa-
tions) signal is an electromagnetic wave of amplitude $\varepsilon_{rf}$, where

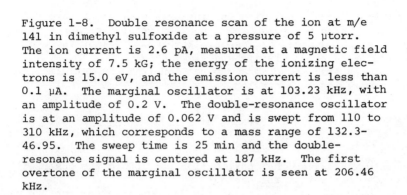

Figure 1-8.  Double resonance scan of the ion at m/e
141 in dimethyl sulfoxide at a pressure of 5 µtorr.
The ion current is 2.6 pA, measured at a magnetic field
intensity of 7.5 kG; the energy of the ionizing electrons is 15.0 eV, and the emission current is less than
0.1 µA.  The marginal oscillator is at 103.23 kHz, with
an amplitude of 0.2 V.  The double-resonance oscillator
is at an amplitude of 0.062 V and is swept from 110 to
310 kHz, which corresponds to a mass range of 132.3-
46.95.  The sweep time is 25 min and the double-
resonance signal is centered at 187 kHz.  The first
overtone of the marginal oscillator is seen at 206.46
kHz.

$\varepsilon_{rf}$ is the electric field intensity at the crest of the wave.
In order to treat the interaction of this wave with moving
charges, it is resolved into two contrarotating circular components, each of which has an amplitude $\varepsilon_{rf}/2$.  The ion in
resonance will follow one of these components in its circular
motion.  This is the essence of resonance--the ion has the same
frequency as the radiofrequency field.  Such an ion will move
subject to an electric field intensity of $\varepsilon_{rf}/2$ as long as it
remains in resonance.  The force (field intensity times charge)
on the ion will be

$$F = ma = \left(\frac{\varepsilon_{rf}}{2}\right)e \qquad (4)$$

from which

$$a = \frac{\varepsilon_{rf}e}{2m} \qquad (5)$$

The velocity of the ion after it has been in resonance for time
t is

$$v = \int_0^t a\ dt = \frac{\varepsilon_{rf}et}{2m} \qquad (6)$$

The instantaneous power A is the product of the force on the ion
and its velocity:

$$A(t) = Fv = \frac{\varepsilon_{rf}^2 e^2 t}{4m} \tag{7}$$

This is the basic equation for instantaneous power absorption. Our derivation of it is based on electrostatic rather than the usual electrodynamic arguments. Any loss of rigor is more than offset by gains in clarity and simplicity.

The average power drawn by an ion from the radiofrequency field while in the detector for a time $\tau$ ($0 < t < \tau$) is

$$A(\tau) = \overline{A(t)} = \frac{1}{\tau} \int_0^\tau A(t) \ dt = \frac{\varepsilon_{rf}^2 e^2 \tau}{8m} \tag{8}$$

Equation (8) gives the contribution per ion to the observed power absorption at resonance.

The power absorbed by an ion depends inversely on its mass. This mass dependence is due to the way in which $F = ma$ enters into Equations (4) and (5).

The translational energy of an ion that has been in resonance for a time $t$ is

$$E_{tr} = \int_0^t A(t) \ dt = \frac{\varepsilon_{rf}^2 e^2 t^2}{8m} + \frac{3}{2} kT \tag{9}$$

The first term is the energy absorbed from $\varepsilon_{rf}$ and the second term is the thermal translational energy, the constant of integration. At room temperature its value is $3.87 \times 10^{-2}$ eV. As an illustration the energy of the $CO_2^+$ ion (mass = $7.3 \times 10^{-26}$ kg) will be calculated assuming that it spends $10^{-3}$ s in the resonance region of a "flat" cell, for which the distance between the upper and lower plates is 1.27 cm. The amplitude of the marginal oscillator wave on the detector plates is 100 mV.

$$\varepsilon_{rf} = \frac{10^{-1} \ V}{1.27 \times 10^{-2} \ m} \simeq 8 \ \frac{V}{m}$$

$$E_{tr} = \frac{8^2 \ V^2 \ m^{-2} \times 1 \ e \times 1.6 \times 10^{-19} \ C \times 10^{-6} \ s^2}{8 \times 7.3 \times 10^{-26} \ kg}$$

$$E_{tr} = 18 \ eV^{25}$$

The energy contributed by the second term in Equation (9) is completely negligible under these conditions. The ion is accelerated to 10 times its thermal translational energy in only

$1.5 \times 10^{-4}$ s.

In general, the resonant ion will be out of phase with the radiofrequency field, and this adds to Equation (9) a term[C20]

$$\frac{\varepsilon_{rf} e v_0 t \cos \gamma}{2}$$

where $v_0$ is the magnitude of the initial velocity and $\gamma$ is the phase angle. Because of the range of the cosine function the extreme values of this term are $\pm \varepsilon_{rf} e v_0 (t/2)$. Replacing $v_0$ by the appropriate velocity from kinetic theory gives for the variation in translational energy due to the phase angle

$$\frac{\varepsilon_{rf} e t}{2} \left(\frac{3 \ kT}{m}\right)^{1/2}$$

Reusing the values of the parameters for which the translational energy is calculated above, the contribution of this term is

$$\frac{8 \ V \ m^{-1} \times 1 \ e \times 10^{-3} \ s}{2} \left(\frac{3 \times 1.38 \times 10^{-23} \ J \ deg^{-1} \times 300 \ deg}{7.3 \times 10^{-26} \ kg}\right)^{1/2}$$

$$= \pm 1.6 \ eV$$

The orbital radius of a resonant ion can be obtained from Equation (6). For a rotating system, the velocity, angular frequency, and radius are related by $v = \omega r$, so that

$$r = \frac{\varepsilon_{rf} e t}{2 m \omega} = \frac{\varepsilon_{rf} e t}{4 \pi m \nu}$$

Equally well,

$$r = \frac{\varepsilon_{rf} t}{2B} \tag{10}$$

by Equation (1). The radius of a $CO_2^+$ ion in resonance at $B = 1.0$ T is, with $\varepsilon_{rf}$ and t as above,

$$r = \frac{8 \ V \ m^{-1} \times 10^{-3} \ s}{2 \times 1.0 \ T} = 0.4 \ cm$$

This value of the radius of the ion trajectory is not much less than half the distance between the upper and lower plates of the "flat" cell. It is thus not difficult to bring ions into

contact with these plates if it is desired to remove them from
resonance.

If $l$ is the length of the resonance region, and $\tau$ the time
the ion requires to drift through it, then Equation (3) for the
drift velocity may be set equal to $l/\tau$. From this comes
$B = \epsilon_{dr}(\tau/l)$. Equation (10) can then be altered to give the
diameter D of an ionic trajectory as that ion leaves the
resonance region of the cell:

$$D = \frac{\epsilon_{rf}l}{\epsilon_{dr}} = \frac{V_{rf}l}{V_{dr}} \tag{11}$$

$V_{rf}$ is the amplitude of the detecting radiofrequency wave; it
was used above to calculate $\epsilon_{rf}$. $V_{dr}$ is the drift voltage in
the resonance region.

Reactions (6), (10), and (11) underestimate their respective quantities of interest to the extent that they ignore the
contribution of 3 kT/2 to the initial value of the velocity,
radius, or diameter. As was shown in the context of the energy
calculation, this error is very small.

If collisions are taken into account all of the above
equations become more complicated. Collisions transfer energy
from ions to neutral molecules so that a steady state is reached
in which ions lose energy to molecules as fast as they gain it
from $\epsilon_{rf}$.[A1] The width of the single-resonance absorption
signals increases with collision frequency; at sufficiently high
pressure this pressure broadening causes a serious problem.
Filament lifetimes decrease with increasing operating pressure,
which is another reason for operation in the µtorr region.

Except when an ion is being irradiated, its kinetic energy
is not much greater than thermal. The magnetic field affects
cyclotron motion by changing the direction, not the speed, of
the ions, whose drift velocity is a small fraction of the room-
temperature speeds of light ions. Virtually no momentum, thus
no kinetic energy, is transferred to an ion as it is ionized by
electron impact *if ionization is the only process*.[26] Electron-
impact ionization normally causes some electronic excitation.[27]
The number of such ions depends on the ionizing energy. Ions
with no more than thermal energy can be produced in charge-
exchange reactions ($M^+ + N \rightarrow M + N^+$, in which M and N are not
necessarily different).[E14]


SIGNAL INTENSITIES

The ICR detection system depends on the power drawn by the ions

from a circuit in which the upper and lower plates of the
resonance region form a capacitor. The number of ions of a
given m/e and the power absorption per ion depend in general on
the location of the ions in the resonance region. Because the
ionic drift velocity is known from Equation (3), distance-
dependent variables can be written as time-dependent variables.[28] The signal intensity[29] is thus

$$I_i = \int A_i(t) \ P_i(t') \ dt \qquad (12)$$

in which $P_i(t')$ is the current of all ions of type i, and $A_i(t)$
is the power drawn from the circuit per ion. Different symbols
for the time dependence of $A_i$ and $P_i$ are used because the former
has t = 0 at the entrance to the analyzer, while the latter has
t = 0 at the electron beam.

The signal intensity as defined here refers to the height
of the signal when the ions are exactly in resonance; it is not
the integrated intensity across the resonance signal. It will
reflect a change in either the number of ions per unit time or
the power absorbed per ion.

For a nonreactive ion formed at the electron beam, the
number of such ions will not vary across the resonance region.
In this case the signal intensity becomes, by Equation (9),

$$I_i = P_i \int_0^\tau A_i(t) \ dt = P_i \ E_{tr,i}(\tau)$$

$$I_i = \frac{P_i \varepsilon_{rf}^2 e^2 \tau^2}{8m_i} \qquad (13)$$

in which the thermal contribution to the translational energy is
ignored.

Knowledge of the relative numbers of each kind of ion in
the cell is essential to a quantitative interpretation of ICR
signal intensities. The question is whether signals resulting
from a sweep of the magnetic flux density B are proportional
only to the numbers of ions of each kind present, or whether
another factor must be taken into account. Because the ion
current $P_i$ is a measure of the number of ions of type i,
Equation (13) should be rewritten as an explicit function of $P_i$:

$$P_i = \frac{8m_i I_i}{\varepsilon_{rf}^2 e^2 \tau^2} \qquad (14)$$

It appears that the ion current is proportional to the product
of the signal intensity and the mass of that ion, but this is a
premature conclusion. Because the signals are produced by
sweeping B, the time dependence of $\tau$ on B must be taken into
account. $\tau$ depends inversely on the drift velocity, $v_{drift}$.
Combining this with Equations (3) and (1) to eliminate $v_{drift}$
shows that $\tau$ depends directly on m when B is swept at constant
$\omega$. Thus the denominator of Equation (14) depends directly on
$m^2$, and numbers proportional to the ion currents of primary ions
are derived by dividing $I_i$ by $m_i$. This is the correction for
mass dependence. It has been extended to the case of double-
resonance signals at high pressure and low conversion.[E44]

ION-TRAPPING CELLS

The drift cell that has been the basis of our discussion thus
far is complemented and in some respects surpassed by ion-
trapping cells. These have their genesis in a short ICR
publication[C12] discussed on page 181 and in the omegatron.[30]
The three ion-trapping cells[31] have greatly extended the

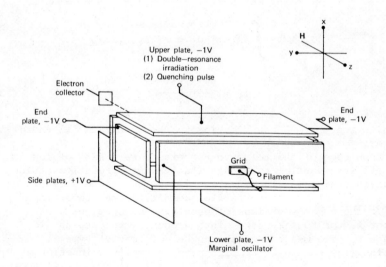

Figure 1-9. The trapped ion cell. Cell dimensions
are 2.54 x 2.54 x 8.9 cm. (Reprinted, by permission,
from *Rev. Sci. Instrum.* 41, 555 (1970), Figure 1.)

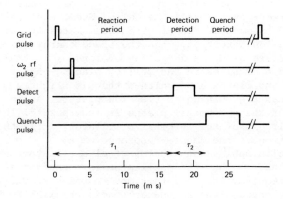

Figure 1-10.  Pulse sequence for PICDR experiments.
(Reprinted, by permission, from *Int. J. Mass Spectrom.
Ion Phys.* 7, 471-483 (1971), Figure 2.)

range of experiments that is convenient via ICR.  They share a
set of important features, notably the time separation of ion
formation and detection.


*The Basic Trapped Ion Cell*[A9,A19]

This one-section cell has no drift field, and the combination of
voltages shown in Figure 1-9 creates equipotentials--paths of
constant potential--that must close on themselves within the
cell in order to keep ions from leaving.  Absorption of radiofrequency energy at the resonant frequency given by Equation (1)
moves the ions into a spiral path centered at the electron beam.
     The unique capabilities of ion-trapping cells derive
largely from their pulsed operation, as shown in Figure 1-10.
Ions are formed, detected, and expelled from the cell by a
series of pulses.  Between formation and detection they are left
to react for a variable time $\tau_1$.  Early in the reaction period
they can be accelerated by a pulse of radiofrequency energy,
shown as $\omega_2$ in the figure.  This mode of operation is known as
"pulsed ion cyclotron double resonance" (PICDR).

The electron beam is controlled by a grid placed between the filament and the opening through which the electrons normally enter the cell. When the grid potential is more negative than the potential on the filament, electrons do not enter the cell. The grid potential is pulsed to a less negative potential for a short time, typically 0.1 ms. This allows ionizing electrons to enter the cell. Use of the voltage pulse on the grid to control the electron beam keeps it out of the cell during ion detection, when the space charge due to the beam would interfere with the detection of ions.

Ions are retained in the cell by placing on the upper, lower, and end plates a DC voltage that is slightly more negative than the voltage on the side plates. As in the drift cell, one of the horizontal plates is incorporated into the resonant circuit of the marginal oscillator, while the other is used to introduce the double-resonance wave.

Detection begins when the marginal oscillator is turned on. In an alternate method the oscillator is turned on at a near-resonant frequency before detection. A pulse of 0.5 V on the side plates changes the resonance condition slightly to match the marginal oscillator frequency.[A6] In double-resonance experiments a pulse-compensator circuit[A19] is required to eliminate interference by $\omega_2$ during ion detection.

The detection period is ended by increasing the voltage on the upper drift plate (quenching pulse). The ions are no longer trapped and move quickly to the cell walls. The (variable) reaction time and the detection time are the same for all ions. The full cycle in Figure 1-10 is repeated on a time base that depends on the desired reaction time $\tau_1$. The various pulses are provided by a commercial time-interval unit. The signals from a number of complete pulse sequences are fed into a boxcar integrator for signal/noise enhancement.[32]

*A Hybrid Trapping and Drift Cell*[A24]

A drift cell can be modified to trap ions. The modifications consist of some wiring changes, the addition of a plate to keep ions from leaving the source in the direction away from the analyzer, and the insertion of two mesh grids, one between the filament and the cell and the other between the electron collector and the cell. The grids are kept at the potential of the adjacent trapping plate. The source region of the cell, in which ions are trapped, is thus protected from the influence of the voltages on the filament and the collector; this enhances ion trapping. While trapping the ions, the source drift plates and the end plate are at ground, the trapping plates at a

positive potential, and the analyzer drift plates are at a
negative potential.

Trapping is ended and analysis begins when all cell volt-
ages are switched to values appropriate to the drift cell.
Trapping and detecting voltages on the drift plates are shown in
Figure 1-11.

*A Trapped Ion Cell with Electrometer Detection*[A43]

This version of the trapped cell is illustrated in Figure 1-12.
The shape of the cell provides highly homogeneous electrical
fields that make long trapping times possible. Radiofrequency
irradiation for ion acceleration to larger orbits enters the
cell via two pairs of wires parallel to the plates. There is no
marginal oscillator; instead, the upper and lower cell plates
are connected to an electrometer. Resonant ions absorb enough
energy to come into contact with these plates, and the current
of ions at a given value of m/e is measured. The ion-current
measurement shows no mass dependence, an advantage of this
method of detection. Ion trapping in this cell is excellent;
the number of ions falls by only 20% after 3 s. The mass range

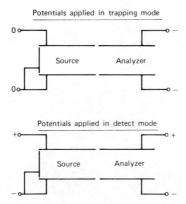

Figure 1-11. Potentials applied to the (hybrid)
trapped ion cell for trapping (above) and detection
(below). (Reprinted, by permission, from *Rev. Sci.
Instrum. 43*, 509 (1972), Figure 2.)

24

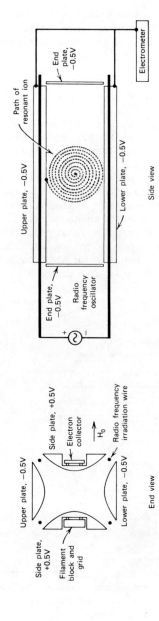

Figure 1-12. End view and side view of an ion-trapping cell with electrometer detection. (Reprinted, by permission, from *Anal. Chem.* 47, 693 (1975), Figure 2.)

extends to at least 500 amu with unit resolution, and calculations indicate that a satisfactory signal can be produced from a sample with a vapor pressure of $10^{-10}$ torr.

In studies of heavy molecules--morphine, acetylsalicylic acid, and the diammonium salt of a nucleotide--the mode of operation is continuous rather than pulsed. The electron beam traverses the cell continuously, and ions are constantly being accelerated to the upper and lower plates by the radiofrequency field. The significance of the chemistry will be presented in Chapter 2.

The cell permits double-resonance experiments via ion ejection (Chapter 5). The ejection signal is carried by the side plates, to which ions of a given m/e move.[A6]

Results obtained in trapped ion cells will be discussed in detail in later chapters. Here we present a brief overview of the variety of experiments engendered by trapped cells. The analysis of data for reaction rates is straightforward because the reaction time can be varied. Figure 1-13[A19] shows the exponential behavior characteristic of pseudo first-order reactions.

The long observation times possible have been skillfully exploited to detect a stable $SF_6^-$ ion at 1.5 x $10^{-7}$ torr.[C69] At this pressure the time between collisions is about 0.2 s, long enough for radiative deexcitation of $SF_6^{-*}$. Experiments in which the anion is observed for much shorter times obscure this mode of stabilization.

Ions can be held sufficiently long to allow forward and back reactions to reach equilibrium.[B15,B25] Measured values of $K_{eq}$ and standard free energies of gaseous acid-base reactions are resolving some of the venerable anomalies of solution chemistry (see Chapter 3).

In the usual ICR experiment in a drift or trapping cell, the neutral molecules are being pumped continuously. No direct knowledge of the product neutrals is available. Neutral products have been observed for an anion-molecule reaction during which the trapped ion cell is not pumped.[A40] The ionic product is monitored until the reaction has gone to a considerable extent. During this phase of the experiment the anions are formed and removed by several cycles of the appropriate pulses. After a time that depends on the reaction rate, ions are quenched for the last time and the electron beam is turned on to ionize the accumulated neutrals and obtain a conventional mass spectrum. The reactant anions are formed in an electron beam that is not sufficiently energetic to produce cations. Thus the product neutrals are not ionized until the electron beam is turned on at a higher voltage for their analysis.

In a test of ICR theory, a trapping cell has provided

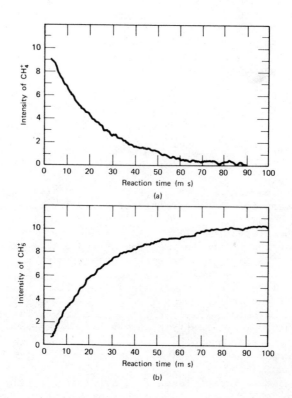

Figure 1-13.   Reactant and product ions as a function
of time for the reaction $CH_4^{+\cdot} + CH_4 \longrightarrow CH_5^+ + CH_3\cdot$.
(A) $CH_4^{+\cdot}$; (B) $CH_5^+$.   Methane pressure is 1.15 μtorr.
(Reprinted, by permission, from *Int. J. Mass Spectrom.
Ion Phys.* 7, 471-483 (1971), Figure 9.)

precise verification of the inverse relationship between signal
linewidth and detection time.[A9]   The cell is also used in
Fourier transform ICR studies.
     The broad utility of ion trapping is well established.   It
is convenient to the quantitative observation of almost any rate
process involving low-pressure gaseous ions.

THE INSTRUMENT

A block diagram of the basic ICR spectrometer is shown in Figure
1-14.  In Figure 1-15 is shown a newer instrument than the one
corresponding to the block diagram.  Its standard modes of
operation--single and double resonance--are as described for the
original instrument.  It also operates as an ion-trapping
instrument.  Other less important modifications extend some of
the operating parameters and render the operation of the instrument simpler and surer.

FOURIER TRANSFORM ION CYCLOTRON RESONANCE SPECTROMETRY

Most spectrometers are scanning instruments; they disperse the
signals and record them one at a time.  The better they do this,
the higher is their resolution.  This is true whether we think
of ultraviolet (UV)-visible, infrared, nuclear magnetic
resonance (NMR), or electron spin resonance (ESR) spectroscopy.
The signal detector is undiscriminating toward signals at
different frequencies, which must reach it "one at a time" if a

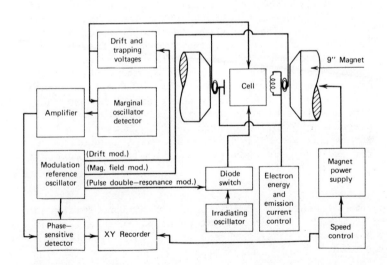

Figure 1-14.  Block diagram of the ICR spectrometer.
(Reprinted, by permission, from *J. Amer. Chem. Soc. 89*,
4570 (1967), Figure 2.)

Figure 1-15. The ICR spectrometer. On the far left
is a vacuum pump used for initial pumping, after which
lower pressures are achieved by an ion pump. Immediately to the right is the magnet unit, which also
includes a dual sample inlet system and the observing
oscillator. At rear center is the magnet power
supply. At the control console on the right the
operator selects or monitors most of the instrumental
parameters. A flat bed recorder is set horizontally
into the left of this console. At its lower right is
a small computer that controls the drift voltages in
the source and analyzer and the pulse sequences needed
for ion-trapping experiments. (Photograph courtesy of
Dynaspec, Inc.)

spectrum--a record of intensity as a function of frequency--is
to be obtained. The energy not detected at a particular time
is wasted. The higher the resolution, the lower the fraction of
energy used.

    For many years the use of a photographic plate or film as a
detector, chiefly in UV-visible and x-ray instruments, was the
only exception to the generalization stated above about
scanning. All of the light falling on the film is being

detected for the entire exposure, so that it is possible to see
very faint spectral lines. The film is, in effect, a multichannel analyzer, capable of accumulating and recording
separately the energy at each frequency.

In conventional ICR spectrometry the resonance frequency is
fixed, and the magnetic field is scanned so as to bring ions of
the different masses into resonance at a time for each kind of
ion that depends on its mass. The method of ion detection is
formally analogous to that of the other scanning instruments.
If, however, a frequency-swept radiofrequency pulse is used to
excite within a few ms all of the ions in the cell, they will be
in resonance simultaneously, and will continue to move in their
cyclotron orbits after the pulse has ended. Their cyclotron
motion will induce a time-dependent voltage in the bottom plate
of the cell, which is connected to a broadband amplifier. The
voltage is the sum of the cyclotron frequencies of all the ions
in the cell, weighted by the number of ions of each mass. This
complex signal is an electromagnetic wave, a function of time
like all such waves. Its amplitude is not directly a function
of any frequency. The wave will decay with time as the ions are
removed from resonance by collisions and is thus a transient.
The experiment is performed on a packet of ions formed in a
trapped ion cell.

Recovery of the frequency information from the wave has
been known in principle since the prize-winning mathematics of
Jean Baptiste Fourier (1810). Let $F(\omega)$ be the amplitude as a
function of frequency and $F(t)$ be the amplitude as a function of
time. The two are related by the Fourier transform[33]

$$F(\omega) = \int_{-\infty}^{\infty} f(t) \exp(-i\omega t) \, dt$$

For any complicated waveform the evaluation of the integral is
extremely tedious, and hence Fourier transform spectroscopy[34]
has become a reality only with the development of computers and
rapid analog-to-digital converters. The first applications were
to infrared and NMR spectroscopy.

A block diagram of the Fourier transform ion cyclotron
resonance spectrometer is given in Figure 1-16. The frequency-
swept radiofrequency pulse is carried by the top plate of the
cell. The signal due to the resonant ions is received on the
bottom plate of the cell and amplified. A second oscillator
provides a wave that is at a higher frequency than any of the
ICR frequencies. This and the signal from the cell are fed into
a mixer, which adds and subtracts each ICR frequency from the
second oscillator frequency. A low-pass filter removes the sum

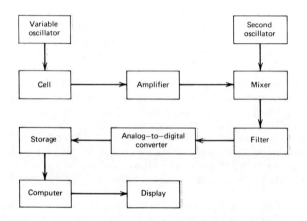

Figure 1-16.  Block diagram of the Fourier transform
ICR spectrometer.

frequencies; the waveform information is now carried by a wave
at a much lower frequency.  This frequency shifting[35]
facilitates subsequent data analysis.

The wave is digitized by an analog-to-digital converter and
stored, and subsequently a computer performs a Fourier transform
on the data.[A36-A39]  The result is $F(\omega)$, a typical ICR scan.
Ions are removed from the cell by a quenching pulse.  Although
useful signals have been obtained from a single packet of ions,
the usual procedure is to store the data from several cycles,
then compute the Fourier transform on the point-by-point
averages of the data.  The result is a better signal/noise
ratio.

Among the attractive features of Fourier transform ICR are
a speed that is much greater than that of magnetic field-sweep
scans, a sensitivity that can be increased simply by averaging
over a greater number of ion formation pulses, and a resolution
that can become truly impressive over a limited region of the

mass range. $CO^+$ and $N_2^+$ have been separated with a resolution of one part in 24 000.[A39]

COLLISIONAL PARAMETERS OF THERMAL MOLECULES

As a point of reference to the familiar collisional parameters
of kinetic theory, we show in Table 1-1 some values of the
number density of molecules, mean free path, and the time
between collisions (reciprocal of the collision frequency) for
$CO_2$ at room temperature. The six pressures easily cover the
operating range of ICR studies. The hard-sphere diameter of $CO_2$
is taken as 4.00 Å.[36]
It is evident from the mean free paths that in the usual
pressure range ($<10^{-4}$ torr) the molecules make most of their
collisions with the cell walls. The necessity of placing the
ions in cyclotron motion in order to achieve a longer path
length is thus confirmed.
At thermal energies, the value of the collision diameter $\sigma$
appropriate for hard-sphere collisions between molecules is
smaller than the value of $\sigma$ for ion-molecule collisions, because
an ion and a molecule are mutually attracted if they are very
close to each other; the attraction between them is stronger
than the Lennard-Jones attraction between molecules. The result
is a decrease in both the mean free path and the time between
collisions for ions. As ion velocities are increased, ion-

Table 1-1

| Pressure | Number Density | Mean Free Path | Time Between Collisions |
|---|---|---|---|
| $P$ | $N' = \dfrac{P}{kT}$ | $\lambda = (2^{1/2}\pi\sigma^2 N')^{-1}$ | $Z_A^{-1} = \dfrac{(mkT)^{1/2}}{4\pi^{1/2}\sigma^2 P}$ |
| *torr* | *molecules/cm$^3$* | *m* | *s* |
| $10^{-2}$ | $3.2 \times 10^{14}$ | $4.4 \times 10^{-3}$ | $1.2 \times 10^{-5}$ |
| $10^{-3}$ | $3.2 \times 10^{13}$ | $4.4 \times 10^{-2}$ | $1.2 \times 10^{-4}$ |
| $10^{-4}$ | $3.2 \times 10^{12}$ | $4.4 \times 10^{-1}$ | $1.2 \times 10^{-3}$ |
| $10^{-5}$ | $3.2 \times 10^{11}$ | $4.4$ | $1.2 \times 10^{-2}$ |
| $10^{-6}$ | $3.2 \times 10^{10}$ | $4.4 \times 10^1$ | $1.2 \times 10^{-1}$ |
| $10^{-7}$ | $3.2 \times 10^9$ | $4.4 \times 10^2$ | $1.2$ |

molecule collisions are also described by a hard-sphere model
(see Table 1-1).

## NOTES

1. Two books that treat solvent effects in some detail are:
   H. A. Laitinen, *Chemical Analysis*, McGraw-Hill, New York
   (1960), Chapter 4 and L. P. Hammett, *Physical Organic
   Chemistry*, 2nd ed., McGraw-Hill, New York (1970), Chapter
   8. One need not read far. Hammett, in the Preface, refers
   to "the discovery that reactions involving bases can go
   $10^{13}$ times faster in dimethylsulfoxide than they do in
   methanol." Solvent effects can also be physical, as in the
   general dependence of organic dye laser wavelengths on the
   solvent.

2. A bibliography on ion-molecule reactions, National Bureau
   of Standards Technical Note 291, U.S. Government Printing
   Office, Washington, D.C. (1966).

3. George A. Sinnott, Bibliography of ion-molecule reaction
   rate data, January 1950 to October 1971, National Bureau of
   Standards Special Publication 381 (1973).

4. See R. W. Kiser, *Introduction to Mass Spectrometry and Its
   Applications*, Prentice-Hall, Englewood Cliffs, N. J.
   (1965), Chapter 2.

5. S. K. Gupta, E. G. Jones, A. G. Harrison, and J. J. Myher,
   *Can. J. Chem. 45*, 3107 (1967).

6. J. H. Futrell and C. D. Miller, *Rev. Sci. Instrum. 37*, 1521
   (1966).

7. J. H. Futrell and T. O. Tiernan, *Science 162*, 415 (1968).

8. First manufactured by Varian Associates, then by Dynaspec,
   Inc. At present no commercial instrument is available.

9. H. Sommer, H. A. Thomas, and J. A. Hipple, *Phys. Rev. 82*,
   697 (1951).

10. We shall use B to denote both the magnetic field and its
    flux density, more commonly and imprecisely called "field
    strength," or simply "field."

11. An ionic charge of unity can be assumed throughout.

12. We make no difference in notation between scalar and vector
    quantities. The sense of the vectors will be indicated
    where needed.

13. The notation in the primary literature does not distinguish
    between $\omega_c$ and $\nu_c$. The latter symbol is never used, even
    though radiofrequency signals are quoted in kilohertz. A
    resonant frequency is calculated for $Ar^+$ by Beauchamp

et al.[E1] as an illustration, but the $2\pi$ is missing and must be supplied in order to repeat the calculation. The terminology is likewise ambiguous; $\omega_c$ is variously called "frequency," "angular frequency," "cyclotron frequency," and "cyclotron resonance frequency." The term "angular velocity," which would be correct and unambiguous for the motion of the ions, is unfortunately never used.

14. The tesla (T) is the MKS unit of magnetic flux density or magnetic induction, and is very convenient for calculations. The gauss (G) or kilogauss (kG) is used in describing the operation of the instrument. 1 tesla = 1 newton/ampere-meter = 10 000 Gauss. The gauss is a part of the CGS system, which is more widely used than the MKS system in the ICR literature.

15. Also referred to as the "analyzer region."

16. R. T. Weidner and R. L. Sells, *Elementary Modern Physics*, 2nd ed., Allyn and Bacon, Boston (1968), p. 338.

17. Also known as the "detecting" or "observing" oscillator.

18. Figure 1-5 shows the ion path in an ion-trapping cell (see page 9). In the drift cell of Figure 1-3, the ion drifts through the cell as it moves in the spiral path.

19. See T. C. O'Haver, *J. Chem. Educ.* 49, A131 (1972).

20. M. Bersohn and J. Baird, *An Introduction to Electron Paramagnetic Resonance*, Benjamin, New York (1966), pp. 205-207.

21. Manufactured by MKS Instruments, Inc. under the trade name "Baratron." The brief description that follows is taken from an MKS design note.

22. This addition of translational energy is often referred to as "ion heating." This is incorrect, because heat is energy that is absorbed by an atom or molecule through all of its variables of motion, subject to the Boltzmann distribution law.

23. Solution reactions are accelerated by an increase in temperature. The opposite can be true for gaseous ion-molecule reactions and depends heavily on the ion translational energy (see R. Wolfgang, *Acc. Chem. Res.* 2, 248 (1969)). The added energy can open new reaction pathways so that other products predominate (see Chapter 4).

24. Double-resonance oscillator and irradiating oscillator are synonymous.

25. Writing the charge once in electrons and once in coulombs and using MKS units elsewhere gives the energy directly in eV.

26. Subsequent fragmentation of the molecular ion can leave the fragment ion with extra kinetic energy. See R. G. Cooks, J. R. Beynon, R. M. Caprioli, and G. R. Lester, *Metastable*

*Ions*, Elsevier, Amsterdam (1973).

27.  R. W. Kiser, *Introduction to Mass Spectrometry and Its Applications*, Prentice-Hall, Englewood Cliffs, N. J. (1965). The energy spread in the electron beam is discussed on pp. 33, 166-167, and 195-198.
28.  This point will be treated in more detail in Appendix 1.
29.  Also known as the ion intensity, signal level, and total power absorption. For the latter, the symbol is usually A (see Appendix 2).
30.  See R. W. Kiser, *Introduction to Mass Spectrometry and Its Applications*, Prentice-Hall, Englewood Cliffs, N. J. (1965), pp. 72 ff. and McIver et al.[A43]
31.  Ion-trapping cell, trapped ion cell, ion-storage cell, and pulsed ion cell are usually synonymous.
32.  See G. M. Hieftje, *An. Chem. 44* (7), 69A (1972).
33.  See, for example, C. K. Mann, T. J. Vickers, and W. M. Gulick, *Instrumental Analysis*, Harper and Row, New York (1974), p. 20.
34.  See G. Horlick, *Fourier Transform Approaches to Spectroscopy, An. Chem. 43* (8), 61A (1971). The article is reprinted in "Instrumentation in analytical chemistry," an ACS Reprint Collection (1973), which also contains articles on Fourier transform IR and NMR spectroscopy.
35.  See J. R. Pierce, *Electrons and Waves*, Science Study Series No. S38, Anchor Books (1964), pp. 184-186.
36.  D. P. Shoemaker and C. W. Garland, *Experiments in Physical Chemistry*, 2nd ed., McGraw-Hill, New York (1967), p. 101.

CHAPTER 2  Reaction Processes

and Their Interpretation


GASEOUS ION-MOLECULE REACTIONS.  A GENERAL DESCRIPTION

Before concentrating on the application of ICR spectrometry to
the study of gaseous ion-molecule reactions (in Chapter 3), we
present a brief review of some general properties of these
reactions.[1]

A typical value for the ionization potential of a molecule
is 10 V.  Thus if a molecule M is ionized to the molecular ion
$M^+$, the ground state of $M^+$ will be 10 eV higher in energy than
the ground state of M.  Conversion of 10 eV into the usual unit
of chemical energy gives 230 kcal/mole.  One tenth of this
energy, or 23 kcal/mole, is a large activation energy, even for
slow reactions in solution.

The hydration energies of ions are of comparable magnitude,
but opposite sign, to the ionization energies of atoms or
molecules.  Thus ions in solution are of much lower energy than
ions in the gas phase.  An example of this is provided by
Figure 1-1.

Activation energies for gaseous ion-molecule reactions are
often immeasurably small and assumed to be near zero; reactions
are very facile.  The energy along the reaction coordinate is
shown for solution and gas-phase reactions in Figure 2-1.  The
gas-phase coordinate is drawn for the case of a reaction with no
activation energy.[2]

Often there is sufficient energy imparted to the ion in its
formation to allow any of several reactions, perhaps breaking
different numbers of bonds.  Several sets of products may result
solely from the reaction of $M^+$ with M.  If the energy of the
ionizing electrons is great enough to produce fragmentation of
$M^+$, the lighter ions can also react with M.  The total number of
reactions that can occur between molecules and ions of the same
substance is usually very much greater than that to which
solution chemists are accustomed.

Despite the greater amounts of energy involved, ion-

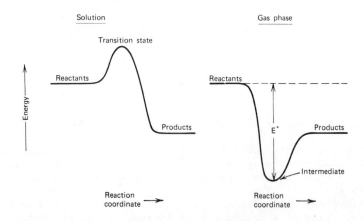

Figure 2-1.   Reaction-coordinate diagram for reactions
in solution and in the gas phase.  Figure courtesy of
J. L. Beauchamp.

molecule reactions can show some of the subtleties of solution
chemistry.  Substantial kinetic isotope effects in the reactions
of $H_2^+$ and $D_2^{+C6}$ and $CH_4^+$ and $CD_4^+$ have been observed,[C5] as well
as a steric effect in a reaction of *exo*- and *endo*-norborneol.[E33]
        The known rate constants of gaseous ion-molecule reactions
generally lie in the range $10^{-8}$-$10^{-12}$ $cm^3$/molecule-s.  Converting the midrange value to solution chemistry units gives
$6 \times 10^{10}$ 1/mole-s, which is comparable to the fastest solution
rate constants.  ($H^+ + OH^- \rightarrow H_2O$ has k ~ $10^{11}$ 1/mole-s.)  This
is not to say that all gaseous ion-molecule reactions are fast.
By present methods it would be difficult to detect a much slower
reaction.
        Radical ions in solution are still the exception rather
than the rule, although they are being increasingly studied.
Neutral free radicals occupy a special place in chemistry but
are not widespread.  In the gas phase the situation is quite
opposite.  Because most molecules have an even number of
electrons, $M^{+\cdot}$ has an unpaired electron; it is a radical cation.
When it reacts with a neutral molecule, one of the products will
be a radical.
        With the exception of $M^{+\cdot}$, the structures of reactant ions

are not always known, except for the case of very simple ions. The structures of product ions are less frequently known; they are inferred in some cases from the subsequent reactions the ions undergo.  Solution chemists have largely been able to abandon this approach to structure determination in favor of physical methods that cannot be applied to gaseous ions.

Mechanisms of simple ion-molecule reactions are being explored with much sophistication and considerable detail in molecular beam studies.  They appear to be of two general types. The direct mechanism is one in which the ion-molecule complex lasts for no more than about $10^{-12}$ s, approximately a rotational period.  If an intermediate is formed at all, it decomposes to products in less time than the complex requires for a rotation. The second type is that of the persistent complex, which lasts considerably longer than a rotation.  There is time for internal redistribution of some of the translational energy of the ion. By the time the complex breaks apart, the fragment that contains the reactant ion departs in a direction that is not influenced by the reactant ion.[3]

TYPICAL REACTIONS

Consider some typical reactions that have been studied in ICR experiments.  In every case the connection between reactant and product ions is known from ICDR results.

$$C_2H_3^+ + C_2H_3Cl \longrightarrow C_4H_5^+ + HCl \qquad (1)$$

$$C_2H_4^{+\cdot} + C_2H_4 \longrightarrow C_3H_5^+ + CH_3\cdot \qquad (2)$$

Reactions (1)[E1] and (2)[E9] are condensation reactions accompanied by the elimination of a neutral product.

$$C_2H_2^{+\cdot} + CH_3CN \longrightarrow CH_3CN^{+\cdot} + C_2H_2 \qquad (3)$$

Reaction (3)[E6] is a charge-exchange reaction, with a $\Delta H$ of +19 kcal, obtained from the difference in ionization potentials of $CH_3CN$ and $C_2H_2$.  The sign of $dk/dE$ is positive; this behavior is typical of exothermic charge-exchange reactions.

$$CH_3^+ + CH_3Br \longrightarrow CH_4 + CH_2Br^+ \qquad (4)$$

Reaction (4)[4] proceeds via a bromonium-ion intermediate; overall it is a hydride-ion transfer.  In Lewis acid-base terms,

$CH_3^+$ is the conjugate acid of the base $CH_4$, and $CH_2Br^+$ is the conjugate acid of the base $CH_3Br$. That the reaction goes in the direction shown establishes $CH_3^+$ as a stronger acid toward hydride than is $CH_2Br^+$.

$$CH_3Br^{+\cdot} + CH_3Br \longrightarrow CH_3BrCH_3^+ + Br\cdot \qquad (5)$$

Reaction (5)[4] illustrates the need for careful mechanistic studies. Without additional information we cannot say if (a) a methyl cation moves from the ionic to the neutral reactant, or (b), a methyl radical moves from the neutral to the ionic reactant. Knowledge of the C-Br bond strengths in $CH_3Br$ and in $CH_3Br^{+\cdot}$ would permit a prediction, but this could be in error because of electronic excitation of the ion.

$$CH_3-\overset{\overset{O}{\|}}{C}-\overset{\overset{O}{\|}}{C}-CH_3^{+\cdot} + CH_3OCH_3 \longrightarrow CH_3-\overset{\overset{O}{\|}}{C}-\underset{+}{O}(CH_3)_2 + CH_3CO\cdot \qquad (6)$$

Reaction (6)[5] is the acetylation of dimethyl ether at its oxygen atom. Similar reactions are observed for simple aliphatic alcohols. These processes may also be considered in terms of Lewis acids and bases; the reaction here illustrates the competition of two bases for the Lewis acid $CH_3CO^+$.

$$BH_4^- + B_2H_6 \longrightarrow B_2H_7^- + BH_3 \qquad (7)$$

Reaction (7)[E5] is also a hydride transfer, but the mechanism is not as simple as it might appear. Double-resonance studies show that the three B atoms are equivalent in the $B_3H_{10}^-$ intermediate.

$$CH_3-\overset{\overset{+OH}{\|}}{C}-H + CH_3-\overset{\overset{OH}{|}}{\underset{\underset{H}{|}}{C}}-CH_3 \longrightarrow CH_3-\overset{\overset{\overset{+}{HOH}}{|}}{\underset{\underset{H}{|}}{C}}-CH_3 + CH_3CHO \qquad (8)$$

$$\Big\downarrow 2\text{-PrOH}$$

$$(CH_3CHOHCH_3)_2H^+ \xrightarrow{\text{2-PrOH}} (CH_3CHOHCH_3)_3H^+$$

$$m/e\ 181$$

Reaction scheme (8)[D4] involves first the protonation of 2-propanol by a fragment ion. This is followed by two reactions in which the newly-formed ion attaches to a molecule of 2-propanol. The final product has a mass of 3M + 1, where M is the molecular mass of 2-propanol. One can only guess at its structure. Protonated polymers (usually dimers) of other oxygen-containing molecules are known. The upper limit of mass detection on some instruments has restricted the search for ionic polymers.

These examples do not exhaust the known reaction types.

With the exception of the second and third steps in Reaction (8), all reactions presented here have a neutral product. The energy released in bond formation between the reactants is taken up to a considerable extent by the energy needed to break a bond and release a neutral species. Only one of the neutral products in Reactions (1) through (8) is mona-tomic; all others can absorb some of the energy by vibration or rotation.

In the absence of a solvent, all of the energy of the reaction must be accounted for by the products. There are few one-product reactions between gaseous ions and molecules, because bond breakage consumes a significant amount of energy. An example of this is provided by the observation that an acetyl cation ($CH_3CO^+$) does not add directly to dimethyl ether to produce the product ion of Reaction (7) at pressures up to $10^{-4}$ torr. The reaction occurs only when there is also a neutral product.

Three-product reactions (one ion and two neutrals) are also uncommon. An example is found on page 108.

The protonated dimer and trimer in Reaction (8) are formed without a concomitant neutral product. In terms of Figure 2-1, they are intermediates that last long enough to be detected (see below). Lacking precise knowledge, we may speculate that their respective potential wells are fairly deep compared to any possible decomposition products. Even if this is not the case, their large numbers of internal degrees of freedom--69 in the case of the dimer--should be able to accommodate a considerable amount of excess energy without concentrating it so as to break a bond.

The chemist who prefers his substances in beakers may be tempted to dismiss some of the more novel product ions as transient curiosities. If an ion is to be detected in an ICR spectrometer, it must exist for at least $10^{-4}$ s in the resonance region of the cell. This is enough time for about $10^9$ vibra-tions per bond. Such an ion coheres through many motions that would sever an insubstantial structure. In contrast, the mean lifetime of a particular $H_3O^+$ ion in strong acid is $10^{-13}$ s.

Further examination of Reactions (1) through (8) shows that the ionic product has an even number of electrons in all cases except the charge-exchange reaction, Reaction (3). An even-electron ionic product would be expected in the reaction of an even-electron ion and molecule; it is usually also the result of a reaction between an odd ion and a neutral molecule.[6]

A very different kind of reaction is the collision-induced fragmentation. The p-chloroethylbenzene molecular ion has been observed in the presence of nitrogen (pressure ratio 1:10).[E4] The molecular ion, $C_8H_9{}^{35}Cl^{+\cdot}$, has a mass of 140, and $C_8H_9{}^{37}Cl^{+\cdot}$ has a mass of 142. Double-resonance irradiation of both 140 and 142 yields an increase in the intensity of the $C_8H_9{}^+$ ion, m/e 105. The differences in mass, 142-105 and 140-105, show that chlorine atoms are lost in the reaction but no nitrogen is incorporated. There is no ion of mass 107. Another fragmentation reaction produces $C_7H_6{}^{35}Cl^{+\cdot}$, m/e 125. Double resonance shows the occurrence of m/e 140→125 but not 142→125. An ion of low concentration at m/e 127 is assumed to come from fragmentation of m/e 142. Again the mass differences are such that nitrogen cannot be incorporated into the products.


KINETIC ANALYSIS OF SINGLE-RESONANCE SIGNALS

The most important aspects of chemical kinetics are the determination of reaction mechanisms and rates. Mechanisms will be treated in Chapter 3 and absolute rates, in Chapter 4. Here we deal briefly with several other aspects of the subject, including relative rates.

Identification of reaction products is straightforward by double resonance in most cases. A check on the results can be made by the older techniques of varying the pressure or ionizing energy. If the ionizing energy is swept through the ionization or appearance potential of a reactant ion, the corresponding product ion will generally appear or disappear along with the reactant, unless the reaction requires that the reactant ion be in an excited state. In this case the appearance potential of the product will be greater than that of the reactant.

If an increase in pressure causes the intensity of the signal of some ion to increase relative to the sum of the signals of all ions, that ion is a product. If the signal decreases, it is that of a primary ion or a product ion that is undergoing further reaction, in which case that signal should show the expected product-ion behavior at a lower pressure. This is shown in Figure 2-2, in which the signal intensities are divided by the respective masses to correct for mass dependence of the marginal oscillator system. Primary ions in the figure are $HCl^{+\cdot}$ and $CH_3F^{+\cdot}$. Secondary ions are $H_2Cl^+$ and $CH_3FH^+$. One of the reactions that produces a secondary ion is

$$HCl^{+\cdot} + CH_3F \longrightarrow CH_3FH^+ + Cl\cdot$$

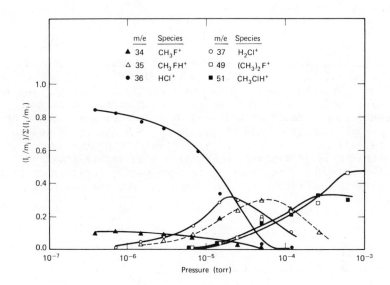

Figure 2-2. Variation of ion densities with pressure for the major ionic species present in a 1:3.5 mixture of $CH_3F$ and HCl at 14.9 eV. The ion densities are normalized signal intensities, corrected for mass dependence. (Reprinted, by permission, from *J. Amer. Chem. Soc. 92*, 7484 (1970), Figure 1.)

The tertiary ions are $(CH_3)_2F^+$ and $CH_3ClH^+$. They are formed by the reactions of secondary ions with neutral molecules. One such reaction is

$$CH_3FH^+ + CH_3F \longrightarrow (CH_3)_2F^+ + HF$$

The question of molecularity is almost trivial in ICR experiments. The primary ions, formed by the unimolecular fragmentation of the molecule-ion, are known from the conventional mass spectrum or can be seen in an ICR scan at very low pressure. The number of three-body collisions occurring in an ICR cell is so small that termolecular reactions are unknown. Nearly all ICR reactions are bimolecular.[C15]

Bimolecularity can be shown in two ways.[E9,E14,C22] Consider the simple case of a single bimolecular reaction between the primary ion $P^+$ and the neutral molecule N to produce the secondary (i.e., product) ion $S^+$:

$$P^+ + N \xrightarrow{k} S^+$$

There will in general also be a neutral product, which can be ignored here. Let P and S be the ion currents of $P^+$ and $S^+$ respectively. In ICR and mass spectrometry the ion current is the usual measure of ion "concentration," whereas molecular concentrations are often given as number densities. Let n be the number density of the molecules N. The number of molecules in the cell at any time is so much greater than the number of ions that n is effectively constant and is mathematically treated as such. The reaction is thus pseudo first-order in the reactant-ion current. The rate equation for the disappearance of $P^+$ is

$$-\frac{dP}{dt} = nkP(t)$$

which gives after integration

$$P(t) = P(0)e^{-nkt}$$

in which P(0) is the rate of ion formation at the electron beam and t the time.

The current of secondary ions is obtained from

$$\frac{dS}{dt} = nkP(t) = nkP(0)e^{-nkt}$$

which integrates to

$$S(t) = P(0)(1 - e^{-nkt})$$

If the ionizing current is constant, P(0) will vary linearly with pressure. Consider the case of a reaction going to a small extent at low pressure. (Pressure is proportional to n, the number density of neutral molecules.) If typical values for n, k, and t are used, the exponent will be $10^{11}$ x $10^{-10}$ x $10^{-3} = 10^{-2}$, so that the exponential can be expanded with retention of only the first two terms:

$$e^{-nkt} \simeq 1 - nkt$$

to a good approximation.  Then

$$S(t) = P(0)nkt$$

Because $P(0)$ depends on $n$, the secondary ion current $S(t)$ will vary as the square of $n$.

Making the same approximation for $P(t)$ gives

$$P(t) \simeq P(0)(1 - nkt)$$

Quantities proportional to $S(t)$ and $P(t)$ are obtained from an ICR spectrum.  Making the substitutions above, the ratio $S(t)/\big(P(t) + S(t)\big)$ reduces to $nkt$, so that a plot of the ratio as a function of the pressure will be linear for a bimolecular reaction.  We cannot, however, obtain a value for the rate constant $k$ by measuring the slope of this ratio as a function of $n$ and supplying an appropriate value of the time $t$.  In the drift cell, ions are being formed continuously in the source, and it is not possible to study what happens to ions formed at some particular zero of time.  Instead, steady-state conditions exist in the cell, in which primary ions are formed, ion-molecule reactions occur, and ions leave the cell continuously. All other things being under control, drift cell spectra do not change with time.

The tests of bimolecularity depend only on the linearity of the data, and not on the slope.  In order to obtain more information, a correction must be made for the mass dependence inherent in ICR detection.[7]  The observed signal intensities are not proportional to the ion currents; this arises because the ions have different residence times in the cell.  Division of the signal intensities of primary ions by the ionic mass gives numbers that are proportional to the ion currents.  For secondary ions the division is by $m^2$.

If a primary ion reacts to form only two products and these are formed by no other ions, the ratio of rate constants can easily be obtained.

$$S_1(t) = P(0)nk_1t_1$$

$$S_2(t) = P(0)nk_2t_2$$

$$\frac{S_1(t)}{S_2(t)} = \frac{k_1t_1}{k_2t_2}$$

$$\frac{k_1}{k_2} = \frac{S_1/t_1}{S_2/t_2}$$

The correction for mass dependence takes into account the differences in ion residence times, and

$$\frac{k_1}{k_2} = \frac{I_1/m_1^2}{I_2/m_2^2}$$

where the I values are the observed signal heights.

## INTERPRETATION OF DOUBLE-RESONANCE SIGNALS

We will first consider the chemical origins of the double-resonance phenomenon. Let k be the rate constant for some reaction and E the kinetic energy of the reactant ion. If $dk/dE$ is positive, the number of product ions will increase as the reactant ion is heated, and $\Delta I$, the change in signal intensity, will be positive, assuming that the sweep-out effect is not a serious problem. If $dk/dE$ is negative, $\Delta I$ will also be negative. The connection between $\Delta I$ and $\Delta H$, the enthalpy change, is shown in the following statements[B1]:

1. A positive $\Delta I$ implies either that $\Delta H$ is positive, or that the reaction has an activation energy that is overcome by heating the ion as shown in Figure 2-3.
2. A negative $\Delta I$ implies that $\Delta H$ is negative.
3. A positive $\Delta H$ implies that $\Delta I$ is positive.
4. A negative $\Delta H$ can appear as either a positive or negative $\Delta I$.

The unambiguous statements (2 and 3) are in agreement with the van't Hoff equation of thermodynamics, even though the reaction systems do not reach equilibrium in the usual ICR studies.

Reactant ions have a distribution of translational energies because they absorb radiofrequency energy for different times before making a reactive collision. They may also differ in their electronic energies as a result of the ionization. The effect of irradiation in this case is to increase (from nearly zero) the number of ions possessing the needed activation energy.

A reaction for which $\Delta I$ is negative must be occurring without the added energy, and must have a $\Delta H$ that is either negative

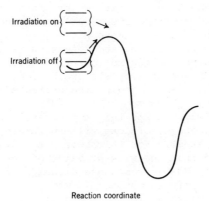

Reaction coordinate

Figure 2-3.   Energies along the reaction coordinate
with and without double-resonance irradiation.


or zero.   Slight exceptions are sometimes postulated when it is
thought that only vibrationally or electronically excited
reactant ions can undergo the reaction.
      Measurement of the total ion current during double-
resonance scans has done much to elucidate the nature of double-
resonance signals.[C21]   The only chemical reason for these
remains the variation of the rate constant with ion energy
($dk/dT$), but total ion current experiments have shown that the
number of available reactant ions can decrease or increase
during a double-resonance experiment.   The sweep-out effect
mentioned in Chapter 1 reduces the ion current by increasing the
cyclotron orbits of resonant ions to such an extent that they
touch the upper and lower plates of the cell, where they are
neutralized.   We see from the equation (Chapter 1, Equation
(10))

$$r = \frac{\varepsilon_{rf} t}{2B}$$

that the radius of resonant ions is greatest at low magnetic
fields.   Light ions will be in resonance at low magnetic fields

for a given detection frequency, and are thus more likely to be swept out of the cell than are heavy ions. The effect is particularly important when $H_2^+$ is a reactant.

The sweep-out effect obviously depends on the distance between the upper and lower plates of the cell. A cell in which this distance is about twice that of the conventional cell has sometimes been used when the sweep-out effect must be minimized. The greater distance between these plates, which carry the drift potentials and the irradiating and detecting radiofrequency fields, results in poorer control over ion movements. It is not unusual to hear ICR spectroscopists speak of a "leaky cell." The ion loss is measurable, and can be corrected for when it is not too great.

The danger always exists that a decrease in signal intensity due to the sweep-out of ions will be interpreted as a negative value of $dk/dT$. For this reason the sign of $dk/dT$ should be obtained from the weakest value of the irradiating field for which the double-resonance signal is clearly visible; the signal in the limit is denoted $(dk/dT)^0$.

A phenomenon known as ion trapping can hold ions close to the electron beam, which decreases the space charge in its immediate vicinity. This effect depends on the total charge in the beam, and is one reason why the emission current must be kept low.

An increase in total ion current is sometimes seen when ions of a prominent m/e come into resonance. This occurs whenever there is considerable loss of ions from the cell, because of an unfortunate choice of cell voltages. Irradiation of a prominent ion can improve the focusing properties so that all ions tend to remain in the cell.

The effects of sweep-out and ion trapping are both seen in Figure 2-4. It shows two conventional ICDR scans with the magnetic field and detecting (observing) frequency fixed, and the irradiating field swept and pulsed. The intensity of the irradiating field for the second scan is twice as great as for the first one. The two lower scans show the total current as a function of frequency under conditions identical to the ICDR scans. At both irradiating field intensities the ion current increases as the resonance frequency $\omega_c$ is approached, then falls in the immediate vicinity of $\omega_c$, indicating that ions are being swept out of the cell as they attain the maximum possible energy from the irradiating field. The ICDR signals are both negative despite the increase in total ion current, which means that $dk/dT$ is negative.

Other factors that influence the appearance of double-resonance signals have not yet been well characterized, but a rough understanding can be gotten from the notion of space

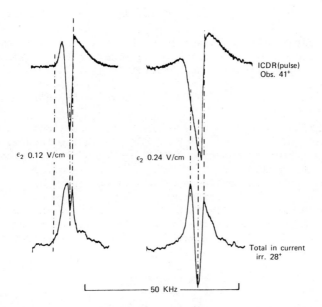

Figure 2-4. Comparison of ICDR and total ion-current scans for the reaction $C_2H_4^{+\cdot} + C_2H_4 \rightarrow C_3H_5^+ + \cdot CH_3$ at two irradiating field strengths. Scans were obtained by sweeping the double-resonance frequency at fixed detector frequency and magnetic field intensity. Double-resonance field strengths are given in V/cm. (Reprinted, by permission, from *Int. J. Mass Spectrom. Ion Phys.* 5, 235 (1970), Figure 5.)

charging, which exists when the electrical fields due to the ions (or electrons at the electron beam) are no longer negligible compared to the external fields, that is, when the ion density is great enough that the ions start to move in response to each other, by Coulomb's law, as well as in response to the external fields. The small but uncontrolled contribution by the ions to the electric and magnetic fields in the cell alters the resonance condition slightly, and is known as "detuning." It is offered as the reason for the asymmetry of the ICDR signals in Figure 2-4.

PROTON AFFINITIES

The negative of the enthalpy change for the reaction

$$M + H^+ \longrightarrow MH^+ \tag{9}$$

carried out in the gas phase is called the "proton affinity" (designated by PA in displayed reactions) of the molecule M. The enthalpy change for the reaction is

$$-PA = \Delta H = \Delta H_f(MH^+) - \Delta H_f(M) - \Delta H_f(H^+) \tag{10}$$

The enthalpy of formation of $H^+$ is 365.48 kcal, and the enthalpies of formation[8] of hundreds of small molecules are known, so the determination of the proton affinity depends on the ability to measure $\Delta H_f(MH^+)$.

Ion cyclotron double-resonance experiments can set narrow limits on the proton affinities of simple molecules.[B1] The measurement is an indirect one based on comparisons, and requires a knowledge of the proton affinities of other molecules. In order to limit the $\Delta H_f$ of $MH^+$, consider the acid-base reaction

$$M_1 + M_2H^+ \longrightarrow M_1H^+ + M_2 \tag{11}$$

for which the reaction enthalpy is

$$\Delta H = \Delta H_f(M_1H^+) + \Delta H_f(M_2) - \Delta H_f(M_1) - \Delta H_f(M_2H^+)$$

Solving for the enthalpy of interest gives

$$\Delta H_f(M_1H^+) = \Delta H - \Delta H_f(M_2) + \Delta H_f(M_1) + \Delta H_f(M_2H^+) \tag{12}$$

The reaction enthalpy is not known, but its sign can be inferred from the variation of the double-resonance signal with irradiating field strength. If the heating of the reactant ion reduces the amount of product (i.e., if $(dk/dE)^0$ is negative; see discussion in the previous section), then the reaction occurs without this input of energy through the double-resonance oscillator, and the $\Delta H$ of the reaction is assumed to be negative or near zero. If $(dk/dE)^0$ is positive, $\Delta H$ can be either positive or negative. The majority of double-resonance signals are negative.

Among the terms on the right in Equation (12), only the first is unknown. If we can say from an ICDR experiment that it is negative, then the sum of the other three terms sets the

*upper* limit to the desired enthalpy of formation.  If another
reaction of the same type has $M_1H^+$ as a *reactant* and gives a
negative double-resonance signal, it will set a *lower* limit to
the enthalpy of formation of $M_1H^+$.  By careful choice of
reactions, quite narrow limits can be set.

The method just described has been used extensively, but
direct measurements of the equilibrium constants of gaseous
acid-base reactions are now routine in the trapped ion cell, and
give more precise results.[B41]

## ANALYTICAL ASPECTS OF ION CYCLOTRON RESONANCE CHEMISTRY

The techniques of analytical chemistry fall into two broad
categories--those in which the sample reacts with some other
reagent, and those in which it does not.  Volumetric analyses
are clearly part of the first group, just as mass spectrometric
analyses belong to the second.  The ICR spectrometer fits both
categories.  At low operating pressures an ordinary mass
spectrum is obtained.  Used in this way, it runs a poor second
to commercial mass spectrometers in terms of resolution, mass
range, and scan speed.  A Fourier transform ICR spectrometer,
however, has high resolution capabilities and a rapid scan time;
it could become competitive with analytical mass spectrometers.
At higher pressures, reactions of molecules with their own ions
or the ions of other molecules occur.

To identify the most likely applications of the technique,
we consider the analytical problems presented by isomerism, for
which the utility of the mass spectrometer is minimal.  This is
especially true for geometrical isomers.  Examination of the
American Petroleum Institute mass spectra of alkenes shows only
minor differences between geometrical isomers.  (Although their
spectra are often sufficiently different from those of their
positional or skeletal isomers to permit identification of the
gross structure of the alkene, characterization of the *cis* or
*trans* nature of the double bond is nearly impossible in most
cases.)  It is possible in principle to distinguish between
hydrocarbon isomers by the use of selective ion-molecule
reactions; studies of the $C_5H_{10}$ isomers suggest that the
reactions of *cis*-2-pentene and *trans*-2-pentene with $C_4H_6^{+\cdot}$ are
quantitatively different.[9]

We may regard $C_4H_6^{+\cdot}$ as a "reagent ion" capable of
distinguishing between similar compounds by their differing
reactivities toward it.  It should be possible to find other
reagent ions, each of which would be analytically useful with
one family of compounds.  Some acylating reagent ions apparently
can distinguish *cis* and *trans* isomers of esters, where the ester

function is locked into axial or equatorial position by the preference of the t-butyl group for the equatorial position.

The difference in reactivity is apparently such that the less hindered equatorial ester reacts faster with the ionic reagent than does the axial.

The ICR spectra of three compounds, all of which have a molecular weight of 54, are shown in Table 2-1.[E11] The mass spectra of the three are quite similar, but their ICR spectra are so different that identification of any one of them from the spectrum of that compound alone is more certain by ICR than by mass spectrometry. The differences between the spectra, and thus between the ion-molecule reaction chemistries, of the two butadienes are particularly striking. The ionizing energy was low enough so that molecular ions were formed without subsequent

Table 2-1

| Ion Formula | M/e | Relative Intensity 1-3 Butadiene | Relative Intensity 1-2 Butadiene | Relative Intensity 1-Butyne |
|---|---|---|---|---|
| $C_7H_9$ | 93 | 100 | 100 | 100 |
| $C_7H_7$ | 91 | 0 | 78 | 60 |
| $C_6H_8$ | 80 | 80 | 6 | 8 |
| $C_6H_7$ | 79 | 100 | 25 | 83 |
| $C_6H_6$ | 78 | 6 | 0 | 2 |
| $C_6H_5$ | 77 | 0 | 5 | 30 |
| $C_5H_7$ | 67 | 27 | 1 | 7 |
| $C_5H_6$ | 66 | 26 | 5 | 8 |
| $C_5H_5$ | 65 | 0 | 0 | 31 |
| $C_4H_7$ | 55 | 6 | 1 | 21 |

fragmentation; they are thus the only reactant ions in the
reactions of the three compounds.

Analysis for specific functional groups is not yet possible
in ICR spectrometry, but the absence of certain groups can be
demonstrated unambiguously in some cases.[E20] Molecules containing a nitrogen or an oxygen atom are easily acylated by the
biacetyl (butanedione) molecular ion or by its chief reaction
product, a triacetyl cation. The reactions that acylate a
molecule M containing N or O can be represented very simply by
letting A· be the acetyl free radical $CH_3CO·$, mass 43. $A_2$ is
thus the biacetyl molecule and $A_3^+$ the very reactive ion formed
from it in the first reaction.

$$A_2^{+·} + A_2 \longrightarrow A_3^+ + A·$$

$$A_2^{+·} + M \longrightarrow MA^+ + A·$$

$$A_3^+ + M \longrightarrow MA^+ + A_2$$

Alkyl halides undergo this reaction to a negligible extent, so
the presence of $MA^+$ excludes a halogen as the functional atom.

The search for reagent ions that react with molecules in
predictable and preferably simple ways is being developed
systematically in studies known collectively as chemical
ionization.[10] A mixture of (usually two) gases is admitted to a
mass spectrometer or ICR spectrometer. The experiment is
easiest when the concentration of the gas that provides the
reagent ion is well in excess of the neutral gas.

The commonest reagent ions are formed from methane in the
reactions

$$CH_4^{+·} + CH_4 \longrightarrow CH_5^+ + CH_3·$$

$$CH_3^+ + CH_4 \longrightarrow C_2H_5^+ + H_2$$

$CH_5^+$ is as familiar to ICR and mass spectrometrists as it is
unexpected to solution chemists. Both ionic products are
Bronsted acids, so that proton-transfer reactions effect
chemical ionization. In other words, $CH_5^+$ and $C_2H_5^+$ chemically
ionize the other gas in a subsequent reaction. If M is the
molecular weight of the gas undergoing chemical ionization, we
can represent some common ions formed as $(M + 1)^+$ by proton
transfer and $(M - 1)^+$ by hydride abstraction or dissociative
proton transfer, and, less frequently, $M^+$, formed by charge
exchange. The latter is important when the reagent is the
cation of a noble gas or nitrogen.

Electron-impact ionization and photoionization are very fast compared to the encounter between a molecule and a relatively massive ionizing agent such as $CH_5^+$. Franck-Condon requirements no longer hold for chemical ionization; in addition, fragmentation is less complex because of lower energy content in the M + 1 ion and the energetics of bond-cleavage processes of even-electron species. The result is usually a few intense peaks rather than a greater number of weaker ones.

Quite obviously, the pressure is great enough that product ions will undergo collisions. Chemical ionization experiments are usually conducted at about 1 torr in mass spectrometers, while in ICR the pressure is less than $10^{-4}$ torr. The ions spend less time in a mass spectrometer source than in an ICR cell, but the time between collisions is much shorter in the mass spectrometer. An ion that will fragment without collisional stabilization is more likely to survive in a mass spectrometer than in an ICR spectrometer. This has been postulated as the reason for the observation of the reaction

$$CH_5^+ + cyclo\text{-}C_6H_{12} \longrightarrow C_5H_9^+ + 2CH_4 \quad (?)$$

in chemical ionization mass spectrometry, but not in ICR.[E39]

Depending on the experimental design, chemical ionization produces information about the ionic chemistry of the reagent ion or about the identity, structure, or reactivity of the neutral molecule. Determination of the molecular weight can be straightforward if the reagent ion is $CH_5^+$ and if the proton affinity of the neutral M is high compared to that of methane. In this case the proton exchange reaction is favored, and produces a prominent $(M + 1)^+$ ion.

We saw above that both $CH_5^+$ and $C_2H_5^+$ are produced from methane. By selectively ejecting one of these ions, an ICR experiment can show unambiguously the contribution of each reagent ion to an ionic product. This has been done for the chemical ionization of some $C_6$ hydrocarbons by $CH_5^+$ and $C_2H_5^+$. The $C_6H_{14}$ isomers consistently lose $H^-$ to form $C_6H_{13}^+$, and $CH_3^-$ to form $C_5H_{11}^+$. Hydride abstraction was due about equally to both reagents, while $CH_3^-$ abstraction was due almost exclusively to reaction with $CH_5^+$. The predominant product in reactions of cyclohexane was $C_6H_{11}^+$, a hydride abstraction, while the predominant reaction of benzene was the addition of a proton to form $C_6H_7^+$.[E39] Thermochemical arguments based on the enthalpies of formation of gaseous ions are frequently used to rationalize such observations.

Chemical ionization of some of the most common esters produced very prominent $(M + 1)^+$ ions and the almost complete lack of $(M - 1)^+$.[E70] This attests to the ease of proton

addition to an ester functional group.  In larger molecules the
formation of $(M - 1)^+$ competes with formation of $(M + 1)^+$.
    A most interesting chemical ionization is[C65]

$$SiH_4 + CH_5^+ \longrightarrow SiH_5^+ + CH_4$$

The silanium ion condenses with ammonia:

$$SiH_5^+ + NH_3 \longrightarrow SiNH_6^+ + H_2$$

An isotopic study of this reaction has provided clues to the
structure of $SiH_5$.
    Mass spectrometrists have made great efforts to extend
their analyses to increasingly heavier--and biologically more
interesting--molecules.  The trapped ICR instrument with
electrometer detection (page 23) has brought this challenging
research area into the domain of ICR chemistry.  In a successful
prototype experiment the diammonium salt of 2'-deoxythymidine
5'-phosphate was introduced into the cell at a sample inlet
temperature of 150°C and attained a pressure of $10^{-6}$ torr with
very slow pumping.  $(CH_3)_3NH^+$ as reagent ion produced an M + 1
ion at m/e 357 and a fragment at M - 114.[A43]
    Even though present theory contributes much to our under-
standing of ion-molecule reactions, a detailed microscopic model
of the chemical processes discussed in this section is not
essential to the development of their analytical applications.
The petroleum industry made extensive use of mass spectrometry
for hydrocarbon analysis long before the fundamental studies of
the physical basis of their fragmentation patterns were carried
out.  Moreover the elegant approaches to the analysis of
fragmentation patterns do not really make the analyst's problem
more straightforward.  It is likely to be the same with ICR;
its applications to practical problems may depend far more on
the reproducibility of experimental conditions and small signals
than on the deeper insights of theory.
    The advantage of ICR over NMR and infrared spectrometry is
that the analysis can be carried out on a smaller sample.


NOTES

1.  Beauchamp[B40] gives an excellent treatment of the subject
    from a different point of view, with a detailed considera-
    tion of the energetics of ion-molecule reactions.
2.  Compare R. Wolfgang, *Acc. Chem. Res. 3*, 48 (1970).

3.  R. Wolfgang, *Acc. Chem. Res. 2*, 248 (1969); *3*, 48 (1970).

4.  J. L. Beauchamp, Ph.D. Dissertation, Harvard University (1968), p. VIII-16.

5.  T. A. Lehman, unpublished results (see Lehman et al.[E27]).

6.  For an interesting comment on this, see the work by Dunbar et al.[E44]

7.  This is discussed in Chapter 1, pages 16 and 20.

8.  All enthalpies refer to the gas phase.

9.  See Chapter 3 for a table of data and fuller discussion.

10. F. H. Field, *Acc. Chem. Res. 1*, 42 (1968); F. H. Field, "Chemical Ionization Mass Spectrometry," in J. L. Franklin (ed.), *Ion-Molecule Reactions*, Vol. 1, Chapter 6, Plenum, New York (1972).

# CHAPTER 3  Reaction Chemistry

## INTRODUCTION

This chapter is not intended to survey all, or even most, of the chemistry of gaseous ions that has been uncovered by ICR techniques. Rather it is an introduction to some aspects of this chemistry that have important implications for the solution chemistry of ions, either because of the interesting analogy that may exist between solution and gaseous behavior or because of the light shed on the importance of solvent effects when there is no analogy.

Accordingly, many of the examples of considerable importance to the understanding of simple collision processes and the reactions of very small molecules have been omitted, since we feel that these may more likely be appreciated by the chemical physicist than by most chemists interested in reactivity and mechanism in solution and in the understanding that a study of reactions in the absence of solvent may bring to these topics.

## STRUCTURE AND REACTIVITY

### Acidity and Basicity

Ion cyclotron resonance is one of several methods that may be used to establish the intrinsic basicity or acidity of unsolvated molecules. Among the others are high-pressure conventional mass spectrometry, in which two molecules compete for a proton in an experiment similar in concept to the ICR experiment,[1] and combination of other heats of formation, in which the proton affinity may be estimated from data from a conventional mass-spectrometric appearance potential and other sources. For $AH^+ \rightarrow A + H^+$, reliable values of $\Delta H_f$ of A and $H^+$ can frequently be obtained from tables. The $\Delta H_f$ of $AH^+$ may be obtained from some fragmentation process observed in a conventional mass spectrometer:

$$AHB + e^- \longrightarrow AH^+ + B + 2e^-$$

These data have been conveniently tabulated as well.[2]  Finally,
*ab initio* calculations have been employed for small molecules.
The results obtained by ICR methods are usually sufficiently
close to those of the other experimental methods that the data
can be interwoven.

It has been difficult to estimate the importance of
solution effects on the relative pK values of acids and bases.
For example, increased branching in the alkyl group of an acid
usually causes a smooth progression in $pK_a$ values, but there are
small inversions in some cases.  It would be convenient to be
able to ascribe this anomalous result to the effect of some
special solvation of the molecule in order to have a totally
consistent picture.  Ion cyclotron resonance offers a method for
obtaining data for relative ranking of compounds on either an
enthalpy (proton affinity) or a free energy (gaseous basicity)
scale.  The proton-affinity measurements depend on the fact that
only exothermic reactions are observed in gaseous ionic chemistry,[3] and on the techniques of double resonance and pressure
plots.

In gaseous basicity measurements, the increase of product
ion intensity in the exothermic direction will be observed.  The
technique is essentially an attempt to allow the various neutral
gases and ions to come to equilibrium by extending the residence
time of the ions in the cell to hundreds of milliseconds in a
trapped ion experiment.  Knowledge of the pressures of neutral
components and measurement of the intensities of the ionic
species allow an equilibrium constant to be calculated.  Under
such conditions many tens of collisions between ions and
neutrals will occur, so that all species are in thermal and
chemical equilibrium.  It should be borne in mind, however, that
in general the number of collisions needed to reach equilibrium
is not known.  One might argue that vibrationally activated
species still exist and therefore distort the value of the
"equilibrium" constant.

The procedure for establishing the proton affinity of a
molecule was outlined in Chapter 2.  Consider in a mixture of
ethylene and water that the reactions

$$C_2H_5^+ + H_2O \longrightarrow C_2H_4 + H_3O^+$$

$$C_2H_4 + H_3O^+ \longrightarrow C_2H_5^+ + H_2O$$

were both found to have a negative $(dk/dE)_0$!  These seem clearly
contradictory results.  However in a mixture of $C_2H_5OH$--in whose

spectrum a peak corresponding to $H_3O^+$ is observed--and $C_2H_4$, no double resonance suggesting the formation of $C_2H_5^+$ from $H_3O^+$ was detected. These results suggest that the proton affinities of $C_2H_4$ and $H_2O$ are fairly similar, but that in a mixture of these compounds, some process or processes occur in which $H_3O^+$ or $C_2H_5^+$ is generated with excess energy, so that the endothermic process is observed. Less complex systems could be found; for example, an upper bound of 171 ± 3 kcal/mole for PA($H_2O$) was obtained by establishing the exothermicity of the reaction

$$CH_2O + H_3O^+ \longrightarrow CH_2OH^+ + H_2O$$

Note that the limit is set because the proton affinity of formaldehyde can be calculated: the $\Delta H_f$ of $CH_2O$ may be obtained from thermochemical tables, the $\Delta H_f$ of $H^+$ may be obtained from spectroscopic data, and the $\Delta H_f$ of $CH_2OH^+$ may be obtained from mass spectrometric studies--for example, data concerning the process $CH_3OH \rightarrow CH_2OH^+ + H + e^-$.

   Likewise a lower bound of 162 ± 2 kcal/mole for PA($H_2O$) was established by observation of the process

$$H_2S^{+\bullet} + H_2O \longrightarrow SH\bullet + H_3O^+$$

since PA(·SH) may be determined from $\Delta H_f(H_2S^{+\bullet})$, $\Delta H_f(H^+)$, and $\Delta H_f(\cdot SH)$. It is interesting to note that by this method proton affinities of radicals as well as neutrals may be determined, or even used as standards; one of the problems in this approach is that there may be some uncertainty in the heat of formation of the radical, which is in fact the case with ·SH.[B1]

   These results were combined to assign a value of 164 ± 4 kcal/mole to the proton affinity of water. Recent results from tandem mass spectrometry give a value of 166 kcal/mole.[4]

*Electronic Effects*

To identify trends in descriptive chemistry, it is sufficient to do a comparative study of the compounds of interest. In this sort of qualitative application of ICR it is not necessary to compare proton affinities with those of known standards. The original study of alkylamines was such a study. In these cases, the proton affinities are quite high, above 210 kcal/mole, and the number of compounds for which proton affinities were available at the time of the study was too small to allow narrow limits to be assigned to the proton-affinity values of the alkylamines.

Note the following series of decreasing proton affinities:[B14]

$$CH_3C(CH_3)_2NH_2 > CH_3C(CH_3)_2CH_2NH_2 > HC(CH_3)_2NH_2 >$$

$$CH_3CH_2CH_2NH_2 > CH_3CH_2NH_2 > CH_3NH_2 > NH_3$$

$$(CH_3CH_2)_2NH > (CH_3)_2NH$$

$$(CH_3CH_2)_3N > (CH_3)_3N$$

$$(CH_3)_3N > (CH_3)_2NH > CH_3NH_2 > NH_3$$

$$(CH_3CH_2)_3N > (CH_3CH_2)_2NH > CH_3CH_2NH_2$$

$$(CH_3)_3N > CH_3C(CH_3)_2NH_2$$

$$(CH_3)_2NH \simeq CH(CH_3)_2NH_2$$

$$(CH_3)_3N \simeq (CH_3CH_2)_2NH$$

Several of these results immediately provide a basis for comment about the anomalous behavior of amines in solution. The dialkylamines show an irregular trend in solution basicities, where

$$CH_3NH_2 < (CH_3)_2NH > (CH_3)_3N$$

$$CH_3CH_2NH_2 < (CH_3CH_2)_2NH > (CH_3CH_2)_3N$$

There had been speculation as to whether this is an intrinsic effect of the structure of the molecule and the ion, or whether it can be explained by peculiar solvation of each member of the series. The results for the amines in the gas phase suggest that enough of the irregularity is ascribable to solvation effects that in its absence an intuitively simple order is restored to the data. A further study of the thermodynamics of solvation of these amines has appeared as another step in the explanation of the order in solution.[B47]

Within the primary amines there is a trend toward increased proton affinity with size, or branching of the alkyl chain. This is reminiscent of the often-quoted inductive effect in solution, as is the trend with increasing substitution by alkyl groups on nitrogen that we have just noticed. Other explanations are possible; for example, substituent effects on rehybridization and polarizability characteristics of the compounds have been examined.[B42] The rehybridization of nitrogen on substitution of H by $CH_3$ is quite susceptible to substituent effects, methyl substitution introducing more s character into the lone pair orbital in amines[5] (but not phosphines). Protonation of the methylamines is accompanied by a rehybridization energy, which is small in the case of $NH_3$ but increases with increasing methyl substitution. This acts in opposition to the stabilization of the ammonium ions by interaction with the polarizable alkyl groups.

The definition of polarizability[6] of a particle is the dipole moment induced by an electric field of unit intensity; polarizability has the dimension of volume. For a molecule in general, polarizability is anisotropic, and the principal polarizabilities (those induced when the field is parallel to one of the semiaxes of the ellipsoid of polarizability) may be described in terms of individual bond polarizabilities. These have been tabulated.[6] It is possible to calculate values of principal polarizabilities from these tables. Some crude correlations can be made by using the concept that polarizability measures how distortable the electrons in a molecule are in the presence of an electric field. For example, the longitudinal polarizability (that is, the polarizability along the bond axis) of a carbon-carbon bond increases from 0.99 $\mathring{A}^3$ for a single bond to 2.80 $\mathring{A}^3$ for a double bond and 3.5 $\mathring{A}^3$ for a triple bond, clearly increasing dramatically as a result of adding deformable pi electrons. Also, for a series of homologs, the polarizability increases as $-CH_2-$ groups are added; the more electrons, the more deformability, other things being equal.

Larger alkyl groups are thus more able to be deformed by, and therefore disperse, the positive charge of the ammonium ion in the substituent, which makes the larger ammonium ion relatively more stable than the smaller one. Recent work has provided more quantitative values of the proton affinity and gaseous basicity of alkylamines.[B23,B24] The various investigators have reported results differing mainly in the assignment of values to methylamine, and in their approaches to calculating $T\Delta S$. Assuming that entropy differences arise from changes in symmetry factors (as suggested by Benson and supported by Kebarle's experimental work) and $T\Delta S$ of $NH_4^+$, 8.6 kcal/mole, one group assigned values of $T\Delta S$ as follows:

|           | $T\Delta S$ |
|-----------|------|
| Primary   | 8.4  |
| Secondary | 8.2  |
| Tertiary  | 7.6  |

The other group adopted the following values:

$$T\Delta S$$

|           |     |
|-----------|-----|
| Primary   | 8.6 |
| Secondary | 8.3 |
| Tertiary  | 7.8 |

The complete data follow in Table 3-1. The increments in gaseous basicity values from each set of workers are generally similar. The more precise data, of course, document the basicity problem all the better. High-pressure work of Kebarle confirms these figures.[7]

To distinguish between the inductive effect and polarizability as the cause of stabilization of larger cations, consider the case of appropriate anions. If the inductive effect of alkyl groups is important, then destabilization of anions by large alkyl groups should occur. If charge dispersal is the explanation, stabilization should occur by large alkyl groups. Experimental results are as follows; the relative acidities of amines occur in the order[B13]

$$(C_2H_5)_2NH > (CH_3)_3CCH_2NH_2 \geq (CH_3)_3CNH_2 \geq (CH_3)_2NH \geq$$

$$(CH_3)_2CHNH_2 > CH_3CH_2NH_2 > CH_3CH_2CH_2NH_2 > CH_3NH_2 > NH_3$$

and, in confirmation of this result, the relative acidities of alcohols occur in the orders[B7]

$$(CH_3)_3CCH_2OH > (CH_3)_3COH > (CH_3)_2CHOH > CH_3CH_2OH > CH_3OH > H_2O$$

$$(CH_3)_3COH \sim n\text{-}C_5H_{11}OH \sim n\text{-}C_4H_9OH > CH_3CH_2CH_2OH > CH_3CH_2OH$$

These results are in sharp contrast to the relative acidities of alcohols in solution, which take a generally opposite trend. It is well known by synthetic chemists that in solution t-butoxide is a stronger base than methoxide, for example; yet in the gaseous phase, the t-butoxide ion has a lower proton affinity than the methoxide ion! The free energy for the reaction $(CH_3O^- + C_2H_5OH = C_2H_5O^- + CH_3OH)$ in the gas phase is $-1.9 \pm 0.2$ kcal/mole.[B32]

These data demonstrate that the relative acidities of alcohols in solution are not the intrinsic acidities, which are measured by the gas-phase reactivity. Solvent effects are of such magnitude as to reverse the apparent behavior of the alcohols. The most important of these effects has been

Table 3-1

| Amine | Gaseous[B23] Basicity | Proton[B23] Affinity | Gaseous[B24] Basicity | Proton[B24] Affinity |
|---|---|---|---|---|
| $NH_3$ | (198±3) | (207±3) | (198.4) | (207±3) |
| $CH_3NH_2$ | (209.8) | (218.4) | 207.9 | 216.3 |
| $C_2H_5NH_2$ | 212.5 | 221.1 | 210.4 | 218.8 |
| $n\text{-}C_3H_7NH_2$ | 213.7 | 222.3 | --- | --- |
| $i\text{-}C_3H_7NH_2$ | 214.7 | 223.3 | 212.8 | 221.2 |
| $n\text{-}C_4H_9NH_2$ | 214.3 | 222.8 | --- | --- |
| $i\text{-}C_4H_9NH_2$ | 214.7 | 223.3 | --- | --- |
| $s\text{-}C_4H_9NH_2$ | 215.8 | 224.4 | --- | --- |
| $t\text{-}C_4H_9NH_2$ | 216.8 | 225.4 | 214.9 | 223.3 |
| $(CH_3)_2NH$ | 216.6 | 224.9 | 213.4 | 222.4 |
| $(C_2H_5)_2NH$ | 221.8 | 230.1 | 219.0 | 227.2 |
| $(n\text{-}C_3H_7)_2NH$ | 226.0 | 234.3 | --- | --- |
| $(i\text{-}C_3H_7)_2NH$ | 223.6 | 231.9 | --- | --- |
| $(n\text{-}C_4H_9)_2NH$ | 224.4 | 232.7 | --- | --- |
| $(CH_3)_3N$ | 221.3 | 229.1 | 218.8 | 226.6 |
| $(C_2H_5)_3N$ | 228.0 | 235.8 | 225.5 | 233.3 |
| $(C_3H_7)_3N$ | 230.3 | 238.2 | --- | --- |
| NH | --- | --- | 216.7 | 224.9 |
| NH | --- | --- | 218.5 | 226.7 |
| NH | --- | --- | 219.8 | 228.0 |

suggested to be steric inhibition of solvation of the bulkier anions, which would of course tend to raise their energies above the energies of the smaller, more extensively solvated anions. The effect of the first solvent molecule on acidity has been studied.  Results are given below for two alkoxide ions.

$$(CH_3O\text{-}H\text{-}OCH_3)^- + C_2H_5OH \longrightarrow (C_2H_5OH\text{-}OCH_3)^- + CH_3OH$$

$$\Delta G = -1.2 \pm 0.2$$

$$(C_2H_5O-H-OCH_3)^- + C_2H_5OH \longrightarrow (C_2H_5OHOC_2H_5)^- + CH_3OH$$

$$\Delta G = -0.4 \pm 0.2$$

The first methanol solvent molecule therefore dampens the difference in free energy between $CH_3O^-$ and $C_2H_5O^-$ from -1.9 to -1.2 kcal, and establishes the trend toward reversal found in fully solvated molecules.[B32]

 These results are clearly at variance with the notion of an intrinsic inductive effect of the alkyl group, an electron-donating effect 'that stabilizes cations and destabilizes anions, as had been proposed on the basis of solution work alone. They are consistent with an interpretation of substituent effects as polarizability-grounded; the larger the group, the more it is able to spread out the effect of a charge of any sign and thus stabilize the ionic species. Similar results were obtained in high-pressure studies of carboxylic acids in the gaseous phase, where butyric was found to be stronger than propionic, and propionic stronger than acetic acid.[8] Compare the anomalies in the solution data for carboxylic acids in Table 3-2.

 Another set of experiments led to an estimate of ranking of free-radical-containing groups with respect to their proton affinities on a scale containing the alkyl groups. The N-nitrosoamines undergo an internal hydrogen abstraction in the gas phase after ionization, because of the lability of the hydrogen attached to the $\beta$ carbon in a sort of McLafferty rearrangement first step.

Table 3-2. Some $pK_a$ Series in
Aqueous Solution That Show Anomalies

| Acid | $pK_a$ |
|---|---|
| $CH_3COOH$ | 4.76 |
| $CH_3CH_2COOH$ | 4.88 |
| $(CH_3)_2CHCOOH$ | 4.86 |
| $(CH_3)_3CCOOH$ | 5.05 |
| | |
| $CH_3COOH$ | 4.76 |
| $CH_3CH_2COOH$ | 4.88 |
| $CH_3(CH_2)_2COOH$ | 4.82 |
| $CH_3(CH_2)_3COOH$ | 4.86 |
| $CH_3(CH_2)_4COOH$ | 4.88 |

$$\begin{array}{c} C_2H_5 \\ \phantom{xxx}\diagdown \\ \phantom{xxxxx}N-N=O^{+\bullet} \\ \phantom{xxx}\diagup \\ C_2H_5 \end{array} \longrightarrow \begin{array}{c} {}^\bullet C_2H_4 \\ \phantom{xxx}\diagdown \\ \phantom{xxxxx}N=N-OH^+ \\ \phantom{xxx}\diagup \\ C_2H_5 \end{array}$$

The ability of this species to transfer a proton to other
dialkyl N-nitrosoamines could then be studied:

$$\begin{array}{c} {}^\bullet C_2H_4 \\ \diagdown \\ \phantom{xx}N=N-OH^+ \\ \diagup \\ C_2H_5 \end{array} + \begin{array}{c} R \\ \diagdown \\ \phantom{xx}N-N=O \\ \diagup \\ R \end{array} \rightleftharpoons \begin{array}{c} R \\ \diagdown \\ \phantom{xx}N^+=N-OH \\ \diagup \\ R \end{array} + \begin{array}{c} {}^\bullet C_2H_4 \\ \diagdown \\ \phantom{xx}N-N=O \\ \diagup \\ C_2H_5 \end{array}$$

In this way it could be demonstrated that the proton affinity of
the species containing $-C_2H_4\bullet$ is bracketed by those of the
dimethyl and dipropyl nitrosoamines, and that the proton
affinity of $(\bullet C_3H_6)(C_3H_7)N-N=O$ (whose conjugate acid is formed
by rearrangement of diisopropyl-N-nitrosoamine, provided the
alkyl group does not rearrange) is bracketed by those of the
diethyl and dibutyl nitrosoamines.  The effect of a $\bullet CH_2CH_2-$
group on proton affinity is therefore fairly similar to that of
$CH_3CH_2$, and that of a $\bullet C_3H_6-$ group is fairly similar to that of
a propyl group.[B9]
   Another measure of reactivity in substituted series is
related to the general behavior of aromatic series of compounds,
with substituents in the *meta* and *para* positions.  Here, for
example, in solution the behavior of substituents for many
classes is so constant through many types of reaction that
substituent constants may be defined describing the relative
electrical effect of the substituent compared with hydrogen.
These are the well-known Hammett substituent constants and their
offspring.  We might have, for example, the relative acidities
of the carboxylic acids $YC_6H_4COOH$, where at $25^\circ$ in water the
p-$NO_2$ substituent increases the acidity of benzoic acid by
decreasing its pK by 0.78 log unit, p-$CF_3$ decreases the pK by
0.54, but p-$CH_3$ and p-$OCH_3$ increase the pK by 0.17 and 0.27
respectively.  The effect has been described as a destabilization of the negatively charged anion by the latter groups'
donation of electrons to the ring, while the former stabilize
the negatively charged anion by attracting electrons toward
themselves.  The mechanism of their action has been debated but
probably involves an electrical effect operating between
preferred positions of the ring through the pi system (resonance
effect) and another effect that does not operate in this fashion
and decreases with distance from the reaction site (field effect
or inductive effect, but these are not synonymous).  In

solution, the operation of these effects in analogous systems
leads to predictability of basicities for the series of pyridines containing these substituents.

The $pK_a$ values of the pyridinium ions are 6.58, 6.03, 5.21,
2.63, and 1.39, for the para substituents $CH_3O$, $CH_3$, $CF_3$, or $NO_2$
respectively; the trend is *opposite* to that of the carboxylic
acids mentioned above because the charge to be stabilized in the
ionic form is of opposite sign.

In the gas phase the order of basicities is similar.
Relative to pyridine, the basicities in the same order differ by
8.0, 5.0, 0.0, -11.0, and -17.0 kcal/mole. Hence to date the
properties of the substituents in pyridine appear to be intrinsic. Moreover these figures allow an estimate of the attenuation of the effect of the substituent on going from the gas
phase to solution. There is an acceptable linear correlation
between the changes in gaseous proton affinity and the changes
in the free-energy change in solution, and the slope of the
correlation line indicates an attenuation by a factor of about
2.1 in the ability of the substituent to influence basicity when
the gaseous ion is placed in aqueous solution.[B22] This factor
has not yet been analyzed, but general medium effects as well as
specific solvation of neutrals and ions will be important.

Closer inspection suggests that stabilization of the
unsolvated pyridinium ion by pi-donor substituents is greater
than in solution.[9] In Figure 3-1, the correlation line is drawn
through points for the nonpi-donor substituents only.

A similar correlation line has been observed for the
relation between the gas-phase acidity of substituted phenols
and their acidity in water. A similar plot shows a correlation
line (for meta and para methyl, fluoro, and chloro substituents)
with a slope of about 4.7. Once again, there is a marked
decrease in the effect of a substituent when the reaction is
studied in water.[B38] Finally, results for the substituted
benzoic acids themselves have been obtained by high-pressure
studies, showing a correlation line with a slope of about 10.[10]

We see, then, that substituent effects follow recognizable
trends, though prediction of these from simple extrapolation of
solution behavior can be misleading. Surprises occur in comparing proton affinities with solution basicities across
functional group divisions, where, for example, one finds the
data shown in Table 3-3.

Note that: (a) isobutylene and the 2-butenes bracket

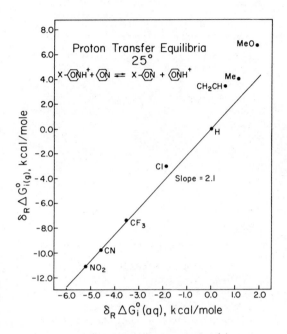

Figure 3-1.  Free-energy changes for proton transfer
between para substituted pyridines in the gas phase
compared with those in aqueous solution.  (Professor
R. W. Taft, Jr., personal communication.)

ethanol and acetone, (b) methanol and ethanol bracket acetaldehyde, and (c) acetaldehyde and acetone bracket ethanol.  The
differences between the intrinsic basicities of oxygen-
containing functional groups are small, and between two homologs
of a given class of compound there will be intervening compounds
of other types.  In solution the alterations of basicity on
substitution are usually not so great.  A result that implies
very great difference in solvation in water is the much stronger
basicity of pyridine relative to ammonia (10 kcal).[B22]  In
aqueous solution ammonia has a $K_b$ four orders of magnitude
larger than the $K_b$ of pyridine.
   There has been an application of proton affinities to the

Table 3-3

| Base | Proton Affinity, kcal/mole[a] |
|---|---|
| $CH_3CH=CH_2$ | 179 |
| $CH_3OH$ | 180 |
| $CH_3CHO$ | 183 |
| $CH_3CH=CH-CH_3$[b] | 183 |
| $C_2H_5OH$ | 186 |
| $CH_3COCH_3$ | 190 |
| $(CH_3)_2C=CH_2$ | 195 |
| $NH_3$ | 207 |
| pyridine | 225 |

[a]A more complete table has recently been published.[11] It includes
results from many laboratories
without comment on the more
acceptable value when there are
significant discrepancies.
[b]*Cis* or *trans*.

problem of nonclassical bicyclic ions.  A proton-affinity
sequence of a variety of ketones and olefins was determined:

$$PA\left( \text{cyclohexanone} \right) \sim PA\left( \text{bicyclic ketone} \right) < PA\left( \text{norbornene} \right) <$$

$$PA\left( \text{bicyclo ketone} \right) \sim PA\left( (CH_3)_2CHOCH(CH_3)_2 \right).$$

$$PA(CH_3OC_2H_5) < PA \left( \text{[bicyclic structure]} \right) < PA(CH_3COC_2H_5)$$

These proton affinities have a number of implications for the chemistry of bicyclic cations. The difference in proton affinity between the bicycloheptanone and the bicyclooctanone is found, upon assigning error limits, to be the same as the difference between cyclopentanone and cyclohexanone; there is no special stabilization for protonated bicycloheptanone. The heat of formation of the bicycloalkyl ions determined from this study, compared with the $\Delta H_f$ for the *saturated* bicyclo hydrocarbons from heats of hydrogenation of the olefins, indicates a 6 kcal/mole stabilization of the bicycloheptyl system over the bicyclooctyl in spite of increased torsional strain and increased nonbonding interactions.[B6] A mechanism of increased delocalization in the bicycloheptyl ion for greater dispersal of the charge was suggested to be of significant importance in the absence of solvation.[12]

For anions changes in sequence of proton affinities are even more striking.

$$\text{[C}_6\text{H}_5\text{]}-O^- < C_2H_5O^- < \text{[C}_6\text{H}_5\text{]}-CH_2^- < CH_3-\text{[C}_6\text{H}_4\text{]}-CH_2^- < OH^-$$

$$HC\equiv C^- < n\text{-}C_4H_9C\equiv C^- < CH_3C\equiv C^- < OH^-$$

In the absence of solvent some carbon acids are stronger than some oxygen acids. Note, too, that in the carbon acids, the effect of alkyl substitution is not the same as the effect of alkyl substitution in oxygen acids. An interplay of inductive and polarizability effects seems to be important. It has been suggested that in the anion any alkyl group is destabilizing relative to hydrogen, but within the anion, the larger alkyl group is more stabilizing than the smaller. Here, then, seems to be evidence for two components of the alkyl-substituent effects, group dipole and polarizability. Suppose that destabilizing group dipoles arise because of the withdrawal, by the unsaturated system from the alkyl system, of charge accumulating near the charge of the anion. Consider withdrawal of charge by an unsaturated substituent from a saturated atom to make this clear. Then the withdrawal is a destabilizing effect relatively unaffected by the size of the alkyl group, but the

size of the alkyl group does affect its polarizability, and
these two effects would oppose each other.[B43]  A similar
ordering has been noted in carboxylic acids and ascribed to the
same interplay of effects.[8]

A particularly interesting point arises here in relation to
rates.  Usually, as was noted in Chapter 2, the exothermic
reaction between an ion and a neutral is very rapid--on the
order of $10^{-9}$ cm$^3$/molecule·s (i.e., $6 \times 10^{11}$/M·s).  Theory calls
for no activation energy in such cases.  Many rates measured are
close to this value.  A few reactions, however, are slow, as in
the case of delocalized anions.

$$CH_3OH + CH_2=CH-CH_2^- \xrightarrow{k_1} CH_3O^- + CH_3-CH=CH_2$$

$$CH_3O^- + \left\langle\!\!\bigcirc\!\!\right\rangle\!-CH_3 \xrightarrow{k_2} CH_3OH + \left\langle\!\!\bigcirc\!\!\right\rangle\!-CH_2^-$$

$$k_1 = 2.5 \pm 0.3 \times 10^{-10} \text{ cm}^3/\text{molecule·s}$$

$$k_2 = 2.0 \pm 0.2 \times 10^{-10} \text{ cm}^3/\text{molecule·s}$$

When two delocalized anions are involved, the rate is particularly slow:

$$\left\langle\!\!\bigcirc\!\!\right\rangle\!-CH_3 + CH_2=CH-CH_2^- \xrightarrow{k_3} \left\langle\!\!\bigcirc\!\!\right\rangle\!-CH_2^- + CH_2=CH-CH_3$$

$$k_3 = 7.5 \pm 0.7 \times 10^{-11} \text{ cm}^3/\text{molecule·s}$$

Thus even though the reaction is exothermic and without competing processes, it might not be observed on the usual milli-
second time scale.  It can, however, be *catalyzed* by the
addition of methanol.  Calculated and observed intensities of
reactant, intermediate, and product ions are in excellent
agreement with each other[C50] (see Figure 3-2).

The question of mechanism arises in these cases.  The fact
that some charge localization must occur in these anions when
the bond to hydrogen is formed suggests a reason for an energy
barrier to bond formation.  Such a barrier does not seem to
exist with acetylide ions, where the charge is localized, since
the rate of proton transfer to acetylide from other species is
*not* outside the range found for simple transfers.  It has been
suggested that an atom carrying lone pairs permits protonation
orthogonal to the pi system, so that delocalization is not an

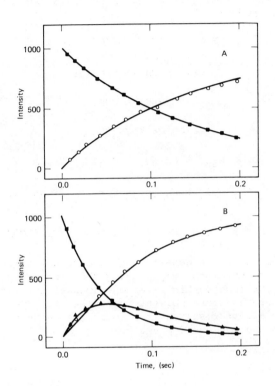

Figure 3-2.  Proton transfer from toluene to the allyl
anion:  (A) uncatalyzed;  (B) catalyzed by methoxide.
Ion intensities versus time:  $C_6H_5CH_2^-$ (o), $C_3H_5^-$ (■),
$CH_3O^-$ (▲).  Pressures in $10^{-6}$ torr for:  (A) propene
7.0, toluene 3.0;  (B) propene 5.1, toluene 3.0,
methanol 1.9.  (Reprinted, by permission, from *J. Amer.
Chem. Soc. 95*, 927 (1973), Figure 1.)


important part of the picture.[C50]
    Related to these ideas of gaseous acidity and basicity is
the concept of solvation ability in the gas phase.  The transfer
of $Cl^-$ between two species, for example,

$$(CH_3NO_2)Cl^- + CH_3OH \rightleftharpoons (CH_3OH)Cl^- + CH_3NO_2$$

may be examined by the same double-resonance techniques used in the acidity and basicity studies. Some solvating power series for chloride are[B34]:

$$CH_3Cl < C_2H_5Cl < (CH_3)_2CHCl$$

$$CH_3F < CH_3Cl \sim CH_3Br$$

$$C_2H_5Cl < CHF_3$$

$$CH_3Cl < CH_3CF_3$$

$$C_2H_5Cl < CH_3NO_2 \sim CH_3CN \sim CH_3OH$$

For alkoxides, the solvating power for the first solvation is[B29]:

$$CH_3OH < C_2H_5OH < (CH_3)_2CHOH < (CH_3)_3COH$$

Both series show a general increase in solvating power as polarizability increases, reproducing the trends found for ionization. Of course this relationship does not last indefinitely as more and more solvent molecules are added; methanol is obviously a better solvent for ionic compounds than is t-butanol in the condensed phase.


*Steric Effects*

In the systems previously studied, no substituents were considered sufficiently large or near the reaction site to interfere with reaction. Indeed, since the reaction is only a proton transfer, it might be difficult to invent a system in which proton transfer might be blocked. Blocking of the transfer of a larger group is more likely.

The general reaction that has been studied most in this regard is the transfer of an acetyl ion. Most commonly this group has been transferred from a biacetyl molecule. The reagent ion is generated as follows[E20]:

$$CH_3COCOCH_3^{+\cdot} + CH_3COCOCH_3 \longrightarrow (CH_3CO)_3^+ + CH_3CO\cdot$$

The approach of this $C_6H_9O_3^+$ complex and transfer of an acetyl group may be so difficult that the reaction is not detected on the ICR time scale. Consider the following series of phenols, under typical conditions in the instrument (pressures of phenols ca. $10^{-6}$ torr, pressure of biacetyl $10^{-5}$ torr, ionizing voltage 30 V).

While the first three react to produce acetylated product, the latter two do not. On the other hand, the analogous compound

does accept an acetyl group under the same conditions.[D10]
    On a more quantitative basis, consider the series of alkylated pyridines

One finds, in competitive studies, the relative rates of acetylation of various members of this series to follow the trend

$$k_H : k_{Me} : k_{Et} : k_{iPr} : k_{tBu} = 1.0 : 1.1 : 0.23 : 0.14 : 0.08$$

This ordering does not correspond to any of the rate theories involving only polarizability and dipole terms, and the decrease for the larger alkyl groups could represent some sort of steric effect.[E64]
    One may look for differences in epimeric aliphatic systems as well. In the set of esters

alkyl groups are fixed into their equatorial and axial positions
by the strong preference of the t-butyl group for the equatorial
conformation.  Their reactivity with a series of acetyl-
transferring species $MCOCH_3^+$ where M is an aldehyde, ketone, or
diketone suggests that the same preference for the equatorial
position is maintained in the gaseous phase and that the generally more accessible equatorial functional group reacts more
readily with most reagents.[13]  Table 3-4 presents rate-constant
ratios of acetyl-transferring ions relative to *cis*- and *trans*-
4-t-butylcyclohexyl acetate.  The ionization energy is ~ 17.5
eV.


*Chemistry of the Gaseous Hydrogen Bond*

In some of the earliest studies of functionally substituted
organic molecules, the appearance of M + 1 ions, the conjugate
acids of the neutral species admitted to the cell, and of 2M + 1
ions, in which two molecules solvate a proton between them, was
noted.  The hydrogen bond between $MH^+$ and M is strong, on the
order of 25 to 40 kcal/mole.  A value of 35 kcal/mole has been


Table 3-4.  Stereoselectivity in Acetyl Transfer
Reactions (see text)

$$MCOCH_3^+ + C_{10}H_{19}OCOCH_3 \longrightarrow C_{10}H_{19}O(COCH_3)_2^+ + M$$

| M | $k_{trans}/k_{cis}$ |
|---|---|
| $CH_3COCOCH_3$ | 1.1 |
| $C_2H_5COCOCH_3$ | 2.6 |
| $CH_3CHO$ | 1.0 |
| $CH_3COCH_3$ | 1.6 |
| $CH_3COC_2H_5$ | 2.0 |
| $CH_3CO-n-C_3H_7$ | neither reacts |
| $CH_3CO-n-C_4H_9$ | neither reacts |

used for thermochemical calculations.   (This was chosen as a
"best" value from high-pressure studies.)

In general, hydrogen bonded ions are observed with protons
between atoms with lone pairs.   No hydrogen bonded species have
yet been observed between delocalized anions, although species
like $(CH_3OH \cdot {}^-OCH_3)$ have been noted above.

The presence of such a bond in a gaseous ion alters the
thermochemistry of possible reactions that might take place in
one of the molecules solvating the proton.   It also changes the
chemistry of the functional group that is part of it.   Much of
the chemistry of alcohols in the gaseous phase is associated
with this phenomenon, and the principal part of this chapter
will review some of the gaseous chemistry of alcohols.   Several
experiments have been done with other functional groups as well.

<u>Condensation and Ionic Dehydration</u>.   In an ICR scan of methanol,
product ions whose elemental formulas are $CH_5O^+$ and $C_2H_7O^+$ are
observed with higher pressure and longer residence time.[E2]
These correspond to protonated methanol and protonated methyl
ether.   At higher pressure ($10^{-5}$ torr) and shorter path lengths
(50 m), other products ($C_2H_9O_2^+$, $C_3H_{11}O_2^+$, and $C_3H_{13}O_3^+$) are
observed.   Because the reactions to form these ions are
exothermic by about 20 kcal/mole, they would carry this energy
as internal excitation, and therefore would have a high
probability of fragmenting unless they were collisionally
stabilized.   Only when the collision time is reduced from $10^{-3}$
to $10^{-6}$ s is the $C_2H_9O_2^+$ observed.   For the same reason, when
stabilized $C_3H_{11}O_2^+$ is accelerated, a collision-induced decomposition of this ion to give $C_2H_7O^+$ is observed.   These observations are consistent with the expectation noted.

The structures postulated for these ions are:

$C_2H_9O_2^+$:

$$\underset{H}{\overset{CH_3}{\diagdown}}O\text{-}\overset{+}{H}\text{-}O\underset{H}{\overset{CH_3}{\diagup}}$$

$C_3H_{11}O_2^+$:

$$\underset{CH_3}{\overset{CH_3}{\diagdown}}O\text{-}\overset{+}{H}\text{-}O\underset{H}{\overset{CH_3}{\diagup}}$$

$C_3H_{13}O_3^+$:

$$\underset{H}{\overset{CH_3}{\diagdown}}O\text{-}\overset{+}{H}\text{-}O\underset{H\text{-}O\diagdown_H}{\overset{CH_3}{\diagup}}{}^{CH_3} \quad \text{or} \quad$$

The structure of the $C_2H_7O^+$ in the run with methanol was pursued
in some detail.  It was confirmed by studying the reactions of
mixtures of methanol and dimethyl ether and of methanol and
ethanol.  The reactivities of the $C_2H_7O^+$ ion in each case were
different; each reacted with $CH_3OH$ to give $C_3H_{11}O_2^+$, but the
complex from ethanol lost water.  That from dimethyl ether did
not, and this behavior was the same as that in methanol alone.
Hence a mechanism for formation of $C_2H_7O^+$ could be the following:

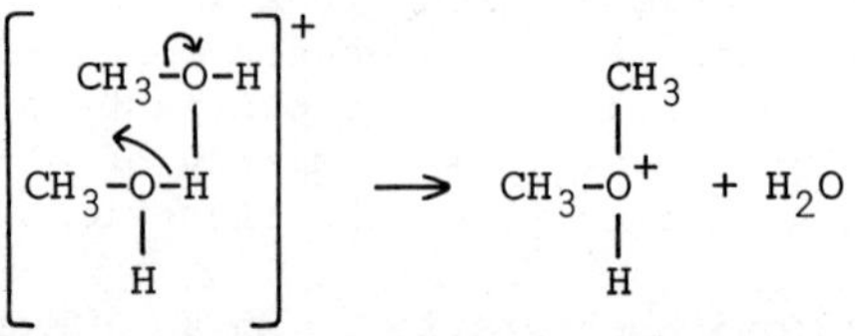

This is an example of one of the two kinds of $H_2O$ expulsion from
an alcohol-solvated proton species.  These have been named
"condensation" and "ionic dehydration."  The reaction in
methanol is a condensation, for which the general mechanism
is[14]:

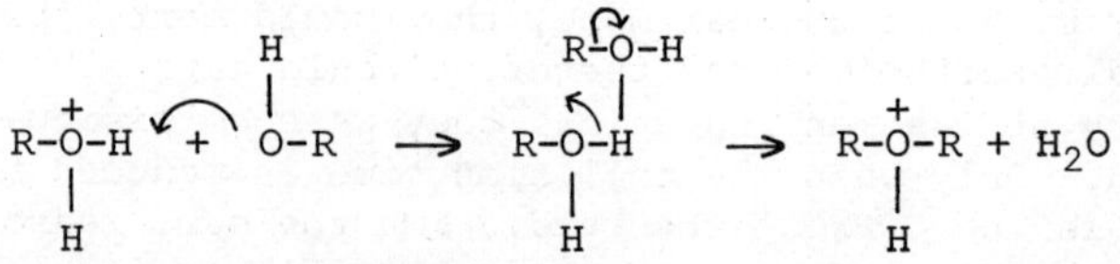

This process is exothermic by about 17 to 24 kcal/mole.  The
molecules solvating the proton need not both be alcohol
molecules; labeling studies have demonstrated a condensation in
a mixed complex as well.[E21]

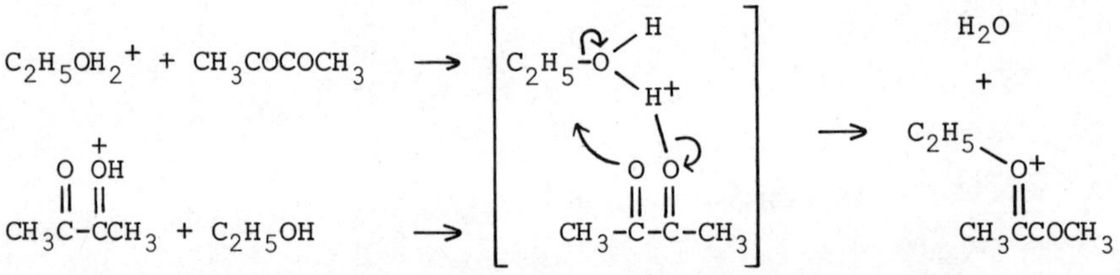

In ionic dehydration, on the other hand, it is the olefin,
rather than water, which is lost as the neutral molecule.[E10]  It
has been observed in isopropanol, cyclopentanol, and t-butanol;
the last example illustrates the mechanism well:

$$\begin{array}{c} CH_3 \quad H \quad H \\ \backslash \quad | \quad + \quad | \\ CH_3-C-O-H-O \quad H \\ / \qquad\qquad | \quad | \\ CH_3 \quad CH_3-C \overset{\curvearrowleft}{=} CH_2 \\ | \\ CH_3 \end{array} \longrightarrow \begin{array}{c} H \quad H. \\ | \quad + \quad | \\ tBu-O-H-O-H \end{array} + C_4H_8$$

Labeling studies in isopropyl alcohol demonstrated a 1,2
elimination, in concert with solution results. The occurrence
of both processes, condensation and ionic dehydration, is
governed by thermodynamic considerations.

Thermochemical Stabilization of Other Cleavage Processes.
Hydrogen bonding often allows reactions to occur in the bound
species that are not seen in their parent ions alone under the
same conditions. A lower limit on the strength of the hydrogen
bond may thus be set. For example:

$$C_2H_5OH^{+\cdot} + CH_3F \longrightarrow CH_2O\overset{+}{-}\overset{}{H}-FCH_3 + CH_3\cdot$$

$$C_2H_5OH^{+\cdot} + H_2O \longrightarrow CH_2O\overset{+}{-}\overset{}{H}-OH_2 + CH_3\cdot$$

or, reversing the initial charge to the nonhydroxylic compound

$$Me_2CO^{+\cdot} + HOCMe_3 \longrightarrow Me_2CO\overset{+}{-}\overset{}{H}-OCMe_2 + CH_3\cdot$$

For these to be observed, the reactions must be exothermic.
In the parent ion of ethanol, the loss of methyl costs 18
kcal/mole, from conventional mass-spectrometry data. Hence the
hydrogen-bond strength must exceed this amount in the cited
proton-bound dimers involving ethanol. It may be as high as 35
kcal/mole. Although the loss of H from ethanol alone costs less
energy (7 kcal), the loss of $CH_3$ is favored here just as in
conventional unimolecular decomposition.

Thermochemistry alone does not predict all aspects of
hydrogen-bond chemistry. Note that the proton affinity of $CH_2O$
is 167; of $H_2O$, 164; and of $CH_3F$, 151. Nevertheless $CH_3CH_2OH$
displaces $CH_2O$, not $H_2O$, from the bound dimer of the two. The
mechanism of the process must involve:

$$\begin{array}{c} H \\ | \curvearrowright \\ C_2H_5O \qquad H-O-H\overset{\curvearrowleft}{-}O=CH_2 \\ | \\ H \end{array}$$

However where such additional features do not complicate the picture, thermochemistry is an adequate guide. In the case of $CH_3F\cdots H\cdots O=CH_2$, it is $CH_3F$ that is displaced by ethanol, as expected on the basis of thermochemistry.[E31]

When intramolecular hydrogen bridges can be formed, they do so in preference to intermolecular bonds. The ethers $CH_3OCH_2CH_2OCH_3$ and $n\text{-}C_8H_{17}OCH_3$ form protonated dimers at nearly the same pressure, but for $CH_3O(CH_2)_5OCH_3$ and $CH_3O(CH_2)_6OCH_3$ no dimer is observed, only monomeric protonated ether. Thus internal solvation has been suggested (see Figure 3-3) as a common phenomenon governing unimolecular decomposition of gaseous ions as well. The large size of the ring system

$$
\begin{array}{ccccc}
CH_3 & & & & CH_3 \\
\backslash & & H & & / \\
& O & + & O & \\
/ & & & & \backslash \\
H_2C & & & & CH_2 \\
\backslash & & & & / \\
& H_2C & & CH_2 & \\
& \backslash & & / & \\
& & CH_2 & &
\end{array}
$$

suggests that a more nearly linear form of the hydrogen bond is more stable than the bent form.[E38] A similar phenomenon, incidentally, has been found in $\alpha,\omega$-diamines through high-pressure studies.[15]

Further examples of reactions operating through proton-bound dimers will appear in the next section.


MECHANISTIC ANALOGIES TO SOLUTION CHEMISTRY


*Nucleophilic Substitution*

The classical example of nucleophilic substitution in solution chemistry is halide exchange in alkyl halides:

$$
\begin{array}{ccc}
R & & R \\
| & & | \\
X^- + R'\text{-}C\text{-}Y & \longrightarrow & X\text{-}C\text{-}R' + Y^- \\
| & & | \\
R'' & & R''
\end{array}
$$

Since most ICR work has been done with positive ions, we should recognize that reactions of the type

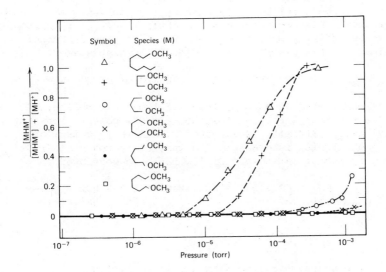

Figure 3-3. Fractional abundances of proton-bound dimers as a function of pressure for methyl n-octyl ether and a series of α,ω-dimethoxyalkanes. (Reprinted, by permission, from *J. Amer. Chem. Soc.* *94*, 3671 (1972), Figure 1.)

$$\begin{array}{ccc} & R & & R \\ & | & & | \\ HX + R'-C-YH^+ & \longrightarrow & H-X-C-R' + HY \\ & | & & | \\ & R'' & & R'' \end{array}$$

are also nucleophilic substitutions. By this definition, the condensation reactions we have discussed in Chapter 2 are nucleophilic substitutions in which the reacting species are held in suitable geometry by the bridging hydrogen. The reaction illustrated may be

considered a displacement of water on a methyl group by
methanol.

The nucleophilic property of attacking groups would not be
expected to parallel basicity exactly, by analogy with solution
chemistry. This may be examined quantitatively, for the concept
of proton affinity may be extended to affinities for other
species.[E18]  For example, the nucleophilic displacement

$$CH_3FH^+ + HCl \longrightarrow CH_3ClH^+ + HF$$

may be considered as a competition of two nucleophiles, HCl and
HF, for the methyl cation, resulting in the observation that the
methyl cation affinity (designated by MCA in formulas) of
hydrogen chloride is greater than that of hydrogen fluoride.  A
series of methyl cation affinities has been developed

|  | $NH_3$ | > | CO | > | $H_2S$ | > | $CH_2O$ | > | HI | > | $H_2O$ | > | HBr | > | HCl | > |
|---|---|---|---|---|---|---|---|---|---|---|---|---|---|---|---|---|
| MCA | 111 | | 82 | | 79 | | 74 | | 67 | | 66 | | 56 | | 51 | |

|  | Zn | ~ | Cd | > | $N_2$ | > | Hg | > | HF | > | $CH_4$ |
|---|---|---|---|---|---|---|---|---|---|---|---|
|  | 45 | | 45 | | 42 | | 38 | | 36 | | 26 |

(The hydrogen halide data are lower limits.)

where the methyl cation affinity is defined as the negative of
the enthalpy of the reaction

$$X + CH_3^+ \longrightarrow XCH_3^+$$

and is expressed in kilocalories per mole.  Note that the
exothermic reaction

$$CH_3FH^+ + N_2 \longrightarrow CH_3N_2^+ + HF$$

fixes nitrogen.  Xenon may also be fixed; the product is
isoelectronic with methyl iodide.[B21]

$$CH_3FH^+ + Xe \longrightarrow CH_3Xe^+ + HF$$

Contrast the order of this series with that for the proton
affinities of the same compounds (data for metal atoms were not
available).  It has been suggested that

|  | $NH_3$ | > | $H_2S$ | > | $CH_2O$ | > | $H_2O$ | > | HI | > | CO | > | HBr | > |
|---|---|---|---|---|---|---|---|---|---|---|---|---|---|---|
| PA | 207 | | 170 | | 165 | | 164 | | 145 | | 143 | | 141 | |

$$HCl > HF > CH_4 > N_2$$
$$140 \quad 137 \quad 126 \quad 116$$

the difference in ordering for different cations may be reflected in the theory of soft and hard acids and bases. Carbon monoxide, then, is relatively more difficult to displace from the soft acid $CH_3^+$ than from the hard $H^+$. A similar shift in reactivity is seen for nitrogen.

A few other reactivity series toward different cations have been noted. Within some organic oxygen bases, the order of nitryl ion affinities (NIA), is[B12]

$$NIA\left((CH_3)_2CHOH\right) > NIA(CH_3CH_2OH) > NIA(CH_3OH) > NIA(H_2O)$$

A few acetyl ion affinities (AIA) have also been studied[E54]:

$$AIA(C_2H_5COC_2H_5) > AIA(C_2H_5COCH_3) > AIA(CH_3COCH_3)$$

Each of these latter trends reflects polarizability and resembles the scale of proton affinities. The first was studied by analysis of mixtures of alcohols and alkyl nitrates,

$$R-\overset{+}{\underset{\underset{H}{|}}{O}}-NO_2 + R'OH \rightleftharpoons R-O-H + NO_2-\overset{+}{\underset{\underset{H}{|}}{O}}-R'$$

and the second in mixtures of ketones.

$$\overset{\overset{+}{O}-COCH_3}{\underset{R-C-R}{||}} + \overset{O}{\underset{R'-C-R'}{||}} \rightleftharpoons \overset{O}{\underset{R-C-R}{||}} + \overset{\overset{+}{O}-COCH_3}{\underset{R'-C-R'}{||}}$$

These processes correspond to transfer of $NO_2^+$ and $CH_3CO^+$ respectively.

Not all nucleophilic displacements occur, even though they may be exothermic. The transfer of a proton through an intermediate involving strong hydrogen bonding must always be considered to be in competition with other processes. For example, in the systems where nucleophilic displacement of HCl by $H_2O$ might be anticipated,

$$CH_3ClH^+ + H_2O \longrightarrow CH_3OH_2^+ + HCl$$

$$C_2H_5ClH^+ + H_2O \longrightarrow C_2H_5OH_2^+ + HCl$$

it was observed only in the latter case. The former system underwent the competitive process

$$CH_3ClH^+ + H_2O \longrightarrow H_3O^+ + CH_3Cl$$

Consider the proton affinities of $H_2O$, $CH_3Cl$, and $C_2H_5Cl$; they are 164, 160, and 167 kcal/mole respectively. Obviously proton transfer from $CH_3Cl$ is exothermic, but from $C_2H_5Cl$ is endothermic. In the $C_2H_5Cl$ system, incidentally, the reaction complex for nucleophilic displacement

$$C_2H_5\text{—}\ddot{\underset{\cdot\cdot}{Cl}}$$
$$\overset{\nwarrow}{H_2O\cdot\cdot\cdot H^+}$$

may be attained either by the previous route or from the reagents $H_3O^+$ and $C_2H_5Cl$. Such reactions, if they are thermodynamically favorable, may provide another channel for decomposition in addition to proton transfer.[E18] A slightly different sort of nucleophilic substitution is a competition of two *cations* for one anion.

$$R_1X + R_2^+ \longrightarrow R_2X + R_1^+$$

One may then speak of the affinity of $R^+$ for $X^-$ or the halide ion affinity. The fluoride ion affinities of the species $CH_nF_{3-n}^+$ were found to be[B35]:

$$CH_3^+ > CF_3^+ > CHF_2^+ > CH_2F^+ > CHFCl^+ >$$

$$CH_3CF_2^+ > CH_3CHF^+ > CH_3CH_2^+$$

Chloride-ion affinities, established analogously, are

$$CF_3^+ > CHF_2^+ > CH_2Cl^+ > CF_2Cl^+ > CHFCl^+ > CHCl_2^+$$

In a few cases here, experiments failed because of slow rates. The reaction

$$CFCl_2^+ + CCl_4 \rightleftharpoons CCl_3^+ + CFCl_3$$

could not be observed in either direction. The hydride affinities are

$$CH_3^+ > CF_3^+ > CH_2F^+ > CHF_2^+$$

and they may be determined by the slow reaction[16]

$$R_1H + R_2^+ \rightleftharpoons R_2H + R_1^+$$

One equilibrium constant for the reactions has been published. In this case the rates of the forward

$$CHF_2^+ + CH_2F_2 \longrightarrow CHF_3 + CH_2F^+ \qquad\qquad K = 0.65$$

and back reactions were determined
($k_f = 1.4 \times 10^{-10}$ and $k_r = 2.2 \times 10^{-10}$ $cm^3$ molecule$^{-1}$ $s^{-1}$) and their ratio gives the same result; hence, conditions leading to equilibrium exist for this reaction on the time scale of the ICR spectrometer. This is an important point, since measured rates may not always be related to reactivities of the lowest energy state.

In view of the fact that few generally consistent trends can be derived from these series on replacement of one halogen atom by another or by hydrogen, the data reflect the opposing effects of electronegativity, pi-donor ability, and polarizability. There are a few consistent trends:

1. Replacing fluorine by chlorine increases the stability, and the fluoride or chloride affinity falls. Polarization stabilization is the dominant effect here.
2. Replacing hydrogen by chlorine does the same.
3. Replacing hydrogen by methyl does the same.

More recently, studies of the type most commonly found in solution have been pursued. These involve displacement of halide from an alkyl halide.[A40,E67]

$$X^- + RY \longrightarrow RX + Y^-$$

The reaction is a backside attack, just as in solution chemistry. The evidence lies in the fact that 1-bromoadamantane and 1-chloroadamantane, whose backside in the reaction center

is protected from attack by the cage structure, are unreactive in these systems, and in the fact that cyclohexyl systems have been shown to react with inversion of configuration by an ingenious experiment in which the neutral products could be analyzed. In attack upon methyl halides, the nucleophilic order $F^- > CH_3S^- > Cl^-$ was found for rates of both methyl chloride and methyl bromide. However the effect of the leaving group depends

on the nucleophile. For the nucleophile $F^-$, chloride is the
better leaving group, but bromide is the better leaving group
when $CH_3S^-$ and $Br^-$ attack the methyl halide. The model
suggested is one in which the rate is fast when the nucleophile
and leaving group have similar properties, although it is not
perfect. Hence this reaction has an activation barrier, as
predicted from theoretical considerations (see Table 3-5).

It is possible that in the gas phase species are not in
their lowest energy states. The flowing afterglow method at
much higher pressures, a technique that permits reduction of ion
energies to thermal distributions, has been applied to nucleophilic displacements,[17] and comparison of the techniques is
definitely necessary. Research by the two methods is converg-
ing.

Another type of nucleophilic attack that has undergone
analysis by many solution chemists is intramolecular nucleophilic attack. In this reaction, and in cases where it inter-
venes as an intermediate step in more complex reactions, the

$$X-(CH_2)_n-CH_2Y \longrightarrow (CH_2)_n \overset{X^+}{\underset{CH_2}{\diagup}} \Big| \quad + \quad Y^- \quad \searrow \text{products}$$

geometry of the attacking group with respect to the leaving
group has played an important role. If one has two isomers,
that which can more readily form a stable ionic intermediate
involving bridging of the substituent will be favored by this

Table 3-5

| X | Y | $k_{reaction}$ $cm^3/molecule \cdot s$ |
|-----|-----|-----|
| F | Cl | $8 \times 10^{-10}$ |
| F | Br | $6 \times 10^{-10}$ |
| Cl | Cl | $0.06 \times 10^{-10}$ |
| Cl | Br | $0.8 \times 10^{-10}$ |
| $CH_3S$ | Cl | $0.8 \times 10^{-10}$ |
| $CH_3S$ | Br | $1.4 \times 10^{-10}$ |

mechanism, and the nature of products or the rate of the
reaction can be used to study the effect.

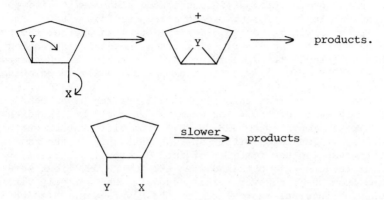

Since the reaction is intramolecular, it might be argued
that, as with other intramolecular processes, analogies could be
studied by conventional low-pressure mass spectrometry. However
reasons for studying the reaction by ion molecule products do
exist:

1.  The gaseous ion molecule product $ZCH_2CH_2-\overset{+}{O}H_2$ formed
    by the reaction
    $ZCH_2CH_2OH^{+\cdot} + ZCH_2CH_2OH \rightarrow ZCH_2CH_2OH_2^{+} + ZC_2H_4O\cdot$
    is more similar to species studied by most solvolysis
    chemists than the unimolecular ionization product
    $Z-CH_2-CH_2-OH^{+\cdot}$.
2.  Ions such as $Z-CH_2-CH_2-OH_2^{+}$ formed by collision can
    be made to have a smaller range of energies than the
    unimolecular ionization product $Z-CH_2-CH_2-OH^{+\cdot}$ formed,
    say, at normal ionizing conditions.

Therefore one can study the decomposition of species with a
narrow energy range and have some confidence that the numbers
obtained roughly represent rate constants applicable to a
definite set of energy conditions, and these results can be
compared to solution reactions studied at a fixed temperature.
        A system very similar to the model system has been the
subject of an ICR study[E52]:

$$YH^{+} + \underset{\underset{OH}{|}}{\overset{\overset{Z}{|}}{CH_2-CH_2}} \rightarrow Y + Z-CH_2-CH_2-\overset{+}{O}H_2{}^{*}$$

$$ZCH_2-CH_2-\overset{+}{O}H_2{}^* \longrightarrow ZC_2H_4{}^+ + H_2O$$

This is the protonation of β-substituted alcohols to give conjugate acids with sufficient excess energy to lose water. The β
substituent Z was varied in order to establish whether the trend
gave evidence for anchimeric assistance, hopefully by analogy
with a well-known trend in solution. The ratio of intensities
$(ZC_2H_4{}^+)/(ZCH_2CH_2OH_2{}^+)$ was taken as a rough guide to relative
rate constants.

Arguments against this simple ratio taking have been
reasonably well refuted. For example, it could be argued that
the energies imparted to $ZCH_2CH_2OH_2{}^+$ from different precursors
in different $ZCH_2CH_2OH$ differ so much that this is the major
factor governing the fraction of ions decomposing. Fairly
similar results were obtained when all $ZCH_2CH_2OH_2{}^+$ ions were
formed from $CH_5{}^+$, however. This suggests that strongly varying
amounts of internal energy are not the key factor. Also while
the ratio changes when ionizing energy is increased from 13 to
70 eV, the order of ratios as a function of Z does not change.

One might also suppose that further reaction of $ZC_2H_4{}^+$
would affect the signal intensity, but there are no ion-molecule
reactions in which this species takes part; the
$(ZC_2H_4{}^+)/(ZCH_2CH_2OH_2{}^+)$ ratio is insensitive to a 50-fold change
in pressure except in the case of the methyl substituent.

A competing reaction, the loss of HZ directly from
$ZCH_2CH_2OH_2{}^+$, is also a problem; HZ is lost if it is thermodynamically more stable than $H_2O$. This reduces the formation of
$ZC_2H_4{}^+$. However a partial correction can be made by adding the
intensity of $C_2H_5O^+$ (i.e., $ZCH_2CH_2OH_2{}^+$ minus HZ) to that of
$ZCH_2CH_2OH_2{}^+$.

The series of substituents in decreasing ability to form
$ZC_2H_4{}^+$, as measured by this ratio, is

$$Br > SH > SCH_3 > OCH_3 \gg Cl, F$$

Thus the reactivity, to the extent that it can be measured this
way, parallels neighboring group reactivity in solution. This
result suggests that anchimeric assistance observed in solution
is fundamentally a property of the substituent and is not a
result of overwhelming solvation effects. It is curious, nevertheless, that the ranking of substituents holds up despite the
fact that in cases of substituents, $Z = OCH_3$, SH, or $SCH_3$, with
a greater proton affinity than OH, one would have expected the
proton to be added to these sites under equilibrium conditions
and that the chemistry of such compounds in the gas phase would
have been significantly altered.

Another similarity is found in comparing the effect of

adding a methyl group to the Z-bearing carbon. In solution this enhances participation by Z; that is, the rate enhancement for loss of $H_2O$ is increased. In the gas phase, it is also larger; for example, the ratio $(CH_3CHBrCH_2{}^+)/(CH_3CHBrCH_2OH_2{}^+)$ is significantly greater than $(BrCH_2CH_2{}^+)/(BrCH_2CH_2OH_2{}^+)$.

Again, the repetition of a trend in the absence of solvent demonstrates that this kind of anchimeric effect in solution is not overwhelmed by solvation of the species involved.

On the other hand, an ICR study of cyclopentanols

indicates that there is only a slight preference for stabilization by the *trans* Z substituent over the *cis* Z substituent.

In textbooks the solution intermediates are pictured as cyclic:

$$\underset{CH_2 \rule{1cm}{0.4pt} CH_2}{\overset{\overset{+}{Z}}{\diagup\diagdown}}$$

The structures of the gaseous $ZC_2H_4{}^+$ ions are of interest. That from $CH_3OCH_2CH_2OH_2{}^+$ reacted similarly to known $CH_3O^+=CH-CH_3$ generated from $CH_3O-CH(CH_3)-CH_2OH$; that from $CH_3SCH_2CH_2OH_2{}^+$ reacted differently from either $CH_3S^+=CH-CH_3$ or $CH_3CH_2S^+=CH_2$ (derived from the appropriate alcohols as in the preceding case) and thus might be cyclic; and that from $BrCH_2-CH_2-OH_2{}^+$ cannot pass through a symmetrical intermediate on further reaction, since only Br from the reactant ion is lost.

$$\underset{CH_2 \rule{1cm}{0.4pt} CH_2}{\overset{\overset{+}{Br}}{\diagup\diagdown}} \xrightarrow[\text{X}]{BrCH_2CH_2OH} \underset{Br}{\overset{H}{\underset{|}{CH_2}}}-CH_2-\overset{|}{\underset{+}{O}}-CH_2-\underset{Br}{\overset{|}{CH_2}} \xrightarrow[-HBr]{\text{loss from either side}}$$

$$CH_3-CH=\overset{+}{O}-CH_2CH_2Br$$

As a result, those $ZC_2H_4{}^+$ ions that enter into reaction with diagnostic reagents are *not* cyclic for Z = OH and Br, but could be for Z = SH.

Finally, there is another type of nucleophilic substitution found in alkyl halides that can be taken as a second example of

an internal nucleophilic substitution. In the methyl and ethyl
halides (with the exception of methyl iodide), the following
reaction is observed.

$$RXH^+ + RX \longrightarrow R_2X^+ + HX$$

The ion formed is considered to arise through a nucleophilic
displacement in a proton-bound activated complex.

This reactivity is similar to that found in very strong acids in
solution. It should be noted that the complex behaves as if it
were symmetrical, and that in principle the charged species and
the neutral species are no longer distinguishable. For the
halogens with two isotopes, chlorine and bromine, this result
was demonstrated as follows.

   Consider the double-resonance experiment in establishing
the pathway from $RXH^+$ to $R_2X^+$. If only the halogen in $RXH^+$
appears in $R_2X^+$, then observation of the $^{37}Cl$-containing product
will show that it comes only from the $^{37}Cl$-containing precursor;
and the $^{35}Cl$-containing product will come only from the $^{35}Cl$-
containing precursor. Take as a second case the possibility
that only the halogen in neutral RX is found in the product
$R_2X^+$. Then, regardless of which isotopic product peak is

1.  $R\underline{Cl}H^+ + RCl \longrightarrow R_2\underline{Cl}^+ + HCl$                  (unique)

2.  $R^{35}ClH^+ + R\underline{Cl} \longrightarrow R_2{}^{35}\underline{Cl}^+$ or $R_2{}^{37}Cl^+ + HCl$   (3:1 ratio)
    $R^{37}ClH^+ + R\underline{Cl} \longrightarrow R_2{}^{35}\underline{Cl}^+$ or $R_2{}^{37}\underline{Cl}^+ + HCl$   (3:1 ratio)

3.  $R^{35}\underline{Cl}H^+ + RCl \longrightarrow R_2{}^{35}\underline{Cl}^+$ or $R_2{}^{37}Cl^+ + HCl$   (equal portions
    $R^{37}\underline{Cl}H^+ + RCl \longrightarrow R_2{}^{35}\underline{Cl}^+$ or $R_2{}^{37}\underline{Cl}^+ + HCl$   of 1 and 2)

irradiated, the isotopic distribution of precursors will be that
found in nature, 3 : 1 for $^{35}Cl$ : $^{37}Cl$. That is, three times
out of four it comes from $^{35}Cl$. For the third possibility, consider the symmetrical intermediate; 50 times out of 100 the
halogen in the product comes from the halogen in the ion and the
other 50 times it comes from the neutral. So we must sum the
previous results, equally weighted, to arrive at the prediction
for this case. For $^{35}Cl$ in the product, seven times out of
eight the $^{35}Cl$ ion is precursor; one time out of eight, $^{37}Cl$.
For $^{37}Cl$ in the product, the $^{37}Cl$ ion is precursor five times
out of eight, $^{35}Cl$, three times out of eight. This last

hypothesis is very close to experimental observation (e.g., $^{35}Cl$: 88/12; $^{37}Cl$: 57/43 for $C_3H_4Cl^+$ formation), so the transition state involves a symmetrical species.[El]

## Electrophilic Substitution

In strongly acidic solution, the nitration of benzene proceeds as shown. Some of the details are omitted.

$$HNO_3 + H^+ \longrightarrow H_2ONO_2^+ \longrightarrow NO_2^+ + H_2O$$

This is an electrophilic aromatic substitution, the replacement of H by $NO_2$ on an aromatic ring. It is a convenient system for comparison of reactivity in solution and gas phases. Ion cyclotron double resonance is especially helpful in sorting out the gas-phase electrophilic substitutions

$$E^+ + C_6H_6 \longrightarrow C_6H_6E^+ \longrightarrow products$$

from other reactions

$$E + C_6H_6^+ \longrightarrow C_6H_6E^+ \longrightarrow products$$

Studies using conventional high-pressure mass spectrometry would yield the origin of the products more reluctantly.

In the gas phase, nitration proceeds by an unusual route confirmed by double resonance. The intermediate is not observed without collisional stabilization because of its high internal energy; it goes on to products. To explain the products,

$$NO_2^+ + C_6H_6 \longrightarrow (C_6H_6NO_2^+) \longrightarrow C_6H_6O^+ + NO$$

there is a loose analogy in solution photochemistry:

$$excess\ NO_2 + C_6H_6 \xrightarrow{h\nu} nitrophenols + NO$$

Other well-known electrophilic aromatic substitutions have different courses after formation of the complex. Recall the Friedel-Crafts reaction:

$$CH_3Cl + AlCl_3 \longrightarrow \overset{\delta+}{CH_3}-\overset{\delta-}{Cl}-AlCl_3$$

$$\overset{\delta-}{AlCl_4}\cdots\overset{\delta+}{CH_3} + \text{⟨benzene⟩} \rightarrow \text{⟨ring}\overset{+}{}\text{⟨}\overset{CH_3}{\underset{H}{}}\text{⟩} + AlCl_4^- \rightarrow$$

$$\text{⟨ring⟩}-CH_3 + HCl + AlCl_3$$

In the gas phase we have

$$CH_3^+ + C_6H_6 \longrightarrow (C_7H_9^+) \longrightarrow C_7H_7^+ + H_2$$

a route more thermochemically favored than one in which a free proton is generated.

In solution, too, the Blanc chloromethylation reaction is an appropriate model for gas-phase reactivity.

$$CH_2=O + HCl \longrightarrow ClCH_2OH \xrightarrow{H^+} ClCH_2OH_2^+$$

$ClCH_2OH_2^+$ behaves as if $ClCH_2^+$ substitutes the ring:

$$ClCH_2^+ + \text{⟨benzene⟩} \rightarrow \underset{ClCH_2}{\overset{H}{}}\text{⟨ring}\overset{+}{}\text{⟩} \rightarrow ClCH_2-\text{⟨ring⟩} + H^+$$

However, the gas-phase process is more like the previous example:

$$ClCH_2^+ + C_6H_6 \longrightarrow (C_7H_8Cl^+) \longrightarrow C_7H_7^+ + HCl$$

A less common solution process is the Hoesch synthesis of acetophenone imine:

$$CH_3C\equiv N + H^+ \longrightarrow CH_3C\equiv NH^+$$

$$CH_3-C\equiv NH^+ + \text{⟨benzene⟩} \rightarrow \underset{H}{\overset{NH}{CH_3-C}}\text{⟨ring}\overset{+}{}\text{⟩} \rightarrow \overset{NH}{CH_3-C}-\text{⟨ring⟩} + H^+$$

In the gas phase, the $CH_3CNH^+$ ion also performs an electrophilic attack, but the subsequent route again tries to avoid formation of the $H^+$ ion.

$$CH_3CNH^+ + C_6H_6 \longrightarrow (C_8H_{10}N^+) \longrightarrow C_7H_8N^+ + H_2$$

Labeling experiments to determine the source of hydrogen cyanide expelled when $CH_2CN^+$ reacts with benzene did not point clearly to a single mechanism. Fifty-seven per cent of the hydrogen cyanide eliminated contained deuterium when $CD_2CN^+$ attacked $C_6H_6$, and 48% contained deuterium when $CH_2CN^+$ attacked $C_6D_6$. This corresponds to neither a specific loss of any hydrogen atom nor complete randomization of hydrogen, so partial scrambling must occur.

No experiments were conducted in any of the preceding studies to establish whether the same path as followed in solution in each of these reactions also occurs in the gas phase. The ionic product would have been $H^+$, which is of very high energy so that other routes would take its place; double-resonance experiments involving a product of m/e 1 have inherent mass-range problems and practical tuning problems. Additionally, the observation of the double resonance signal for the origin of so ubiquitous an ion does not necessarily define the path. Collisionally induced processes would also be responsible for it, possibly even from the same precursor.

It is not surprising to find in these systems that at low pressure the intermediate complex cannot be detected. Reaction complexes frequently require stabilization by collision with a third body before they are observed. On the other hand, one can detect such intermediates if one produces them by *transfer* of $E^+$ instead of *addition*:

$$A-E^+ + B \longrightarrow A + BE^+$$

In such a case the overall complex dissociates to the ion we wish to observe and a neutral fragment.

For example, the intermediate nitration product is detected when $NO_2^+$ is transferred:

$$CH_2O-NO_2^+ + \langle\bigcirc\rangle \longrightarrow C_6H_6NO_2^+ + CH_2O$$

and the Friedel-Crafts reaction intermediate can be observed in the case of toluene (benzene itself gives an ion of very low intensity).[E17]

$$(CH_3CO)_2^+ + \langle\bigcirc\rangle-CH_3 \longrightarrow C_9H_{11}O^+ + CH_3CO$$

Some substituent effects have been studied for attack on neutral benzenes, with surprising results. One of the problems with such systems is that the atom attacked by the electrophile is not specified. Hence any relative reactivity shown by a molecule is the reactivity of all of its positions. In much of the work described heretofore the functional group was clearly defined and the chemistry observed could be described in terms of functional group reactivity. In the case of substituted benzenes, one can establish neither whether the reactions are occurring on the substituent or at a ring atom, nor, for that matter, whether there is generally more than one site of reaction in the ring. Note the data observed below (Table 3-6) for the transfer of $NO_2^+$ from $CH_2O$ to substituted benzenes.

$$CH_2ONO_2^+ + C_6H_5Y \xrightarrow{k} CH_2O + C_6H_5YNO_2^+$$

One expects that for an electrophilic attack the rate of reaction will increase as the substituent becomes more electron-donating; that is, as its Hammett sigma constant decreases. The withdrawing nitro substituent ($\sigma = +0.78$) slows attack by $E^+$ on the ring because ring electrons are less available for attack.

drift of electrons

(no attractive interaction with $E^+$)

For an electron donating substituent such as $NH_2$ ($\sigma = -0.33$), the rate of attack is increased.

Table 3-6. Rate Constants (relative to benzene) for Transfer of $NO_2^+$

Increasing Hammett $\sigma$ Constant
Decreasing Electron Donation
←

| Y | $NO_2$ | $CF_3$ | Cl | F | H | $CH_3$ | $C_2H_5$ | $OCH_3$ | $NH_2$ |
|---|--------|--------|------|------|-----|--------|----------|---------|--------|
| k | 10 | 0.3 | 0.25 | 0.55 | 1.0 | 0.3 | 0.0 | 0.0 | 0.0 |

This explains the order of reactivity in nitration in solution, but not the reversal of the order of nitration rates in the gas phase. In the gas phase, the more the substituent donates electrons, the slower it reacts with $CH_2ONO_2^+$. Even if we consider that reaction at the functional group might be involved in the case of nitrobenzene, it is still difficult to explain why such normally reactive species as $C_6H_5NH_2$ and $C_6H_6OCH_3$ are unreactive. What has been suggested is that the transfer of $NO_2^+$ to aromatic systems may not be electrophilic in all characteristics. If the reaction complex is formed rapidly and if $NO_2^+$ transfer is slow, it could acquire the characteristic of a

nucleophilic attack within the complex if, for example, the oxygen atom were the one forming the bond to the aromatic compound. Hence the reactivity with substituents would reflect this, the slowest step of the reaction sequence. Theoretical studies indicate that oxygen attack of $NO_2^+$ on $C_2H_4$ is one of three possible routes for reaction.[18]

On the other hand, the reactivity in acetylation forms the pattern more normally expected from a model of electrophilic attack. Only a few compounds are acetylated by $CH_3COCOCH_3^+$; most are acetylated by the triacetyl ion formed from biacetyl, $(CH_3CO)_3^+$. It becomes clear in each series that electron-donating substituents accelerate the reaction, while electron-withdrawing substituents decelerate it below the level of detectability in the instrument. The exception, $NO_2$, is almost certainly a poor analogy because of reaction on the nitro group; consider that nitromethane also reacts with $(CH_3CO)_3^+$. With this exception, which is explainable, the reactions examined so far make sense in general.

On close inspection, however, there is a further problem. Consider the table (Table 3-7) in which relative reactivities toward transfer of acetyl from biacetyl, acetone, and ethyl acetate are compared. There is no great problem in explaining the changing cutoffs for observing acetylation; the three acetyl compounds have different acetyl affinities, and so fit into the table of relative affinities of substituted benzenes at different places. It is curious, however, that above the threshold for reactivity the different acetylating groups operate at considerably different rates. Not only are the relative rates rather different, based on phenol as the base compound, but there are rather serious exchanges of position in the table.

Table 3-7

|          | Source | | |
|----------|--------|--------|--------|
| Aromatic | $CH_3COCOCH_3$ | $CH_3COCH_3$ | $CH_3COOC_2H_5$ |
| $C_6H_5OH$ | 100 | 100[a] | 100[a] |
| $C_6H_5NO_2$ | 40 | 69 | 33 |
| $C_6H_5NH_2$ | 33 | 84 | 20 |
| $C_6H_5OCH_3$ | 29 | 10 | 4 |
| $C_6H_5F$ | 27 | _[b] | - |
| $C_6H_5CH_3$ | 19 | 3 | - |
| $C_6H_5C_2H_5$ | 7 | - | - |
| $C_6H_6$ | 2 | - | - |

[a]Each column is chosen with phenol as the base for that column. The relative rates of phenol with the three acetylating agents were not established.
[b]No product was observed.

This indicates that the reaction complex is not just a simple one in which $CH_3CO^+$ is transferred from donor to acceptor, but instead that there are rather specific interactions within the complex between donor and acceptor, of a nature such that they are not the same for a given benzene as one changes the acetyl donor. Chemical interactions occur, and these interactions affect the rate of transfer to some extent in each case.[19]

In conclusion, then, there are numerous analogies to the pathway for electrophilic attack upon aromatic systems found in solution. In some cases, the complexes formed in solution and the gas phase lead to different ionic and neutral products, because the solvation of the ionic product, a proton, in solution greatly alters the relative energies of possible products. Nevertheless such typical electrophiles as $NO_2^+$, $CH_3^+$, $CH_2Cl^+$, $CH_3CNH^+$, and $CH_3CO^+$ attack aromatic rings. The former, upon closer examination by substituent effects, does not follow the reactivity pattern expected for strict electrophilic attack; the latter does. It is pleasing to see that reactivity of aromatics follows in a general way the patterns learned from solution behavior of substituted compounds, but much work remains to be done in this area.

*Elimination*

The elimination reaction in solution is of the type

$$RCH_2-CH_2-X \longrightarrow R-CH=CH_2 + HX$$

It may be acid-catalyzed

$$R-CH_2-CH_2-OH + H^+ \longrightarrow RCH_2-CH_2-\overset{+}{O}H_2$$

$$R-CH_2-CH_2-OH_2{}^+ \longrightarrow R-CH=CH_2 + H_2O$$

or base-catalyzed

$$Z^- + RCH_2-CH_2-X \longrightarrow H-Z + RCH_2=CH_2 + X^-$$

$$HZ + X^- \longrightarrow Z^- + HX$$

and, depending on the sequence of the loss of H and X from the carbon chain, may be first- or second-order. We have already discussed some examples of elimination reactions in the gas phase by noting alcohol behavior.

The ionic dehydration reaction mentioned as a consequence of hydrogen bonding is an *acid-catalyzed* elimination. Deuterium labeling in isopropanol confirms the process as typical of a 1,2-elimination.

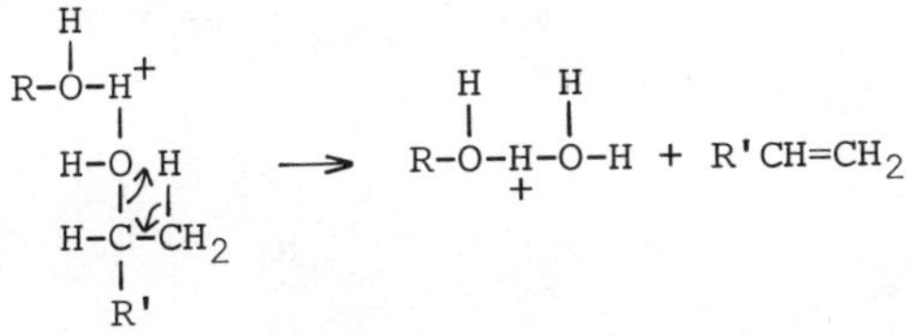

Comparison of appropriate labeling experiments here leads to a value of 1.7 for the isotope effect favoring abstraction of H over D, that is, in isopropanol labeled in one methyl group

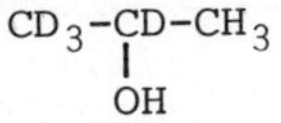

$$CD_3-\underset{\underset{OH}{|}}{CD}-CH_3$$

the neutral $CD_3-CD=CH_2$ is expelled 1.7 times more often than $CH_3-CD=CD_2$. The central hydrogen, as noted above, is not involved in this loss, as experiments with $CD_3CHOHCD_3$ indicate. Information about isotope effects has been thus far insufficient

to allow interpretation of this number as elaborately as would
be possible in the case of a solution reaction. That it is
different from unity indicates, not surprisingly, that changes
in bonding to the C-1 hydrogen occur at the activated complex
for the slowest step, if indeed this is a multistep process.[D4]
     A considerable amount of work has been accomplished on the
course of this reaction in 2-butanol.[E36]

$$CH_3-\overset{\overset{\displaystyle OH}{|}}{\underset{\underset{\displaystyle H}{|}}{C}}-\overset{\overset{\displaystyle H}{|}}{\underset{\underset{\displaystyle CH_3}{|}}{C}}-H$$

The products can be rationalized in terms of various reactant
conformations:

|  14%  |  60%  |  26%  |  ---  |

The alumina-catalyzed reaction gives the products indicated
above, as a condensed phase model. In the gas phase, however,
the products were in the following ratios as gauged from
labeling experiments.

| | |
|---|---|
| Methyl cyclopropane | $\leq 0.18$ |
| 1-Butene | $\geq 0.16$ |
| *trans*-2-Butene | $\geq 0.35$ |
| *cis*-2-Butene | $\geq 0.35$ |

The inequality signs result from uncertainty regarding the size
of the isotope effect for different processes, scrambling of
hydrogen atoms among positions in the chain, and operation of
other processes. Thus the gas-phase reaction appears to be less
sensitive to conformational effects than in solution; witness
the lowered discrimination between structures leading to *cis-*
and *trans*-2-butene, resulting in apparently nearly equal amounts
of these products. The gaseous elimination *requires* a *syn*
configuration of H lost and OH lost,

but this does not introduce any unusual constraints not present
in the solution case. A further analysis of the product
distribution was not made.

A final observation about the 2-butanol system is another
result pertaining to lack of stereospecificity. In this ionic
dehydration both the *threo* and *erythro* forms of the $3\text{-}d_1$ alcohol

eliminate $H_2O$ in preference to HDO by a factor of two. Clearly,
the subtleties of stereochemical probing show that nonbonded
interactions play a smaller role than in solution, at least for
this molecule. We might have anticipated that the preferred
process in solution would lead to a dominant loss of HOD in one
case and $H_2O$ in the other. Here, however, no process is
preferred. In one case, HOD loss comes from formation of *cis*-2-
butene, and in the other, from formation of *trans*-2-butene. The
labeling results suggest that these are equally likely, and
apparently the sum of other processes involving loss of only
H--such as anti-Saytsev elimination and α-elimination--is also
equally likely. Hence the 2:1 ratio exists in the gas phase.

There are also base-catalyzed eliminations. A reported

example is an elimination of the stable HF molecule. The
attacking base is conveniently generated from an alkyl nitrite.
(This is a standard trick for forming alkoxide ions.)

$$CH_3ONO \xrightarrow{e^-} CH_3O^- + NO$$

$$CH_3O^- + CH_3CF_3 \longrightarrow
\begin{bmatrix}
CH_3O^- \\
\vdots \\
\underset{|}{H} \quad \underset{|}{F} \\
CH_2-CF_2
\end{bmatrix}
\longrightarrow
\begin{matrix}
CH_3OHF^- \\
+ \\
CH_2{=}CF_2
\end{matrix}$$

Other products have been found from the reaction, $CF_3CH_2^-$ and
$F^-$. The process illustrated, however, is an E2 elimination in
which the eliminated components, HF, have a *cis* configuration.[E68,E69] The other polyatomic product ion, $CF_3CH_2^-$, corresponds to collapse of the hydrogen-bonded activated complex by
proton transfer, and $F^-$ may result from a number of routes,
perhaps involving internally excited intermediates. These
reactions have been studied for the polyfluoroethanes; the
analogous processes in β-haloethanols and trichloroethane have
also been studied. Elimination of HF is more important than
that of HCl or HBr in describing their respective reactivities.[E68,E69]


*Addition to Multiple Bonds*

At least in principle, addition to olefins in solution is the
reverse of the elimination reaction that we have just studied:

$$\underset{H}{\overset{H}{>}}C{=}C\underset{H}{\overset{H}{<}} \quad + \quad H_2O \quad \longrightarrow \quad CH_3CH_2OH$$

In solution, addition is stepwise. An electrophile, a radical,
or a nucleophile may add in the first step.

$$\underset{H}{\overset{H}{>}}C{=}C\underset{H}{\overset{H}{<}} \quad + \quad H^+ \quad \longrightarrow \quad CH_3CH_2^+ \quad \longrightarrow \quad products$$

$$\underset{H}{\overset{H}{>}}C{=}C\underset{H}{\overset{H}{<}} \quad + \quad R\cdot \quad \longrightarrow \quad RCH_2-CH_2\cdot \quad \longrightarrow \quad products$$

$$\underset{H}{\overset{H}{>}}C=C\underset{H}{\overset{COOC_2H_5}{<}} + CN^- \longrightarrow N\equiv C-CH_2-CH=C\underset{OC_2H_5}{\overset{O^-}{<}} \longrightarrow \text{products}$$

An important application of these reactions is in polymerization of olefins. For example, cationic polymerization of ethylene may be initiated and propagated as follows.

$$\underset{H}{\overset{H}{>}}C=C\underset{H}{\overset{H}{<}} + H^+ \longrightarrow C_2H_5^+$$

$$\underset{H}{\overset{H}{>}}C=C\underset{H}{\overset{H}{<}} + C_2H_5^+ \longrightarrow C_4H_9^+$$

$$\underset{H}{\overset{H}{>}}C=C\underset{H}{\overset{H}{<}} + C_nH_{2n+1}^+ \longrightarrow C_{n+2}H_{2n+5}^+$$

The polymeric cations produced by ion-molecule reactions in gaseous ethylene by attack of different ions upon neutral ethylene offer an analogy in gaseous ion chemistry to these processes. The mechanisms of the processes are not resolved, and although this reaction has been very ably studied from the physical viewpoint, many questions remain regarding the nature of the complexes formed. What evidence is available comes from studies of higher analogs.

$$C_2H_4^{+\cdot} + C_2H_4 \longrightarrow C_3H_5^+ + CH_3\cdot$$

$$C_2H_4^{+\cdot} + C_2H_4 \longrightarrow C_4H_7^+ + H\cdot$$

$$C_3H_5^+ + C_2H_4 \longrightarrow C_5H_7^+ + H_2$$

$$C_3H_5^+ + C_2H_4 \longrightarrow C_5H_9^+$$

Other fragment ions from ethylene undergo similar reactions with neutral ethylene.

$$C_2H_3^+ + C_2H_4 \longrightarrow C_3H_3^+ + CH_4$$

$$C_2H_3^+ + C_2H_4 \longrightarrow C_4H_5^+ + H_2$$

$$C_2H_3^+ + C_2H_4 \longrightarrow C_2H_5^+ + C_2H_2$$

$$C_2H_5^+ + C_2H_4 \longrightarrow C_3H_5^+ + CH_4$$

$$C_2H_2^{+\cdot} + C_2H_4 \longrightarrow C_3H_3^+ + CH_3\cdot$$

$$C_3H_3^+ + C_2H_4 \longrightarrow C_5H_7^+$$

$$C_2H_2^{+\cdot} + C_2H_4 \longrightarrow C_4H_5^+ + H\cdot$$

When only one product is observed, the reaction is reminiscent of the early steps of a cationic polymerization in solution.[E9]

Extensive hydrogen scrambling in these reactions of ethylene tends to obscure the reaction mechanisms. For example, the allyl (or cyclopropyl) ion formed from ionizing a mixture of $C_2H_4$ and $C_2D_4$, or ionizing $CD_2CH_2$, contains almost the same ratio of d, $d_2$, $d_3$, and $d_4$ species, approximately: 0.25 to 1 to 1 to 0.23. The expected ratio for complete scrambling is 0.17 to 1 to 1 to 0.17. This scrambling is typical of that found in many long-lived ion-molecule reaction complexes. This is particularly the case in the small hydrocarbons where the internal energy is high; many bonds are of equal energy, and barriers to exchange of hydrogen between carbon atoms are low.

As the intermediate complex becomes larger, however, less scrambling occurs in ion-molecule reactions. The close analogy with solution reactivity disappears here, however. In allene, for example, data suggest that the four-center complex

$$D_2C=C=CD_2^{+\cdot}$$
$$+$$
$$H_2C=C=CH_2$$
$$\longrightarrow$$
$$\left[\begin{array}{cc} D_2C \!\!\!-\!\!\!-\!\!\!-\!\!\!-\!\!\! C=CD_2 \\ \mid \quad\quad\quad \mid \\ H_2C \!\!\!-\!\!\!-\!\!\!-\!\!\!-\!\!\! C=CH_2 \end{array}\right]^{+\cdot}$$

for the loss of ethylene decomposes faster than scrambling occurs, so that $C_2H_4$ is lost three times more frequently than a complete scrambling mechanism would predict. A more complex mechanism is necessary for loss of ethylene from propyne.

$$D_3C-C\equiv C-D$$
$$\text{HC}\equiv C-CH_2 \quad H \quad \longrightarrow \quad C_4D_3H^+ + C_2H_3D$$

Here the amount of $C_2H_3D$ lost is about twice as much as would be expected after complete scrambling. Similarly, the loss of acetylene from propyne

is more likely to be as $C_2HD$ than as $C_2H_2$ or $C_2D_2$. Total scrambling of hydrogens requires that 50% of acetylene lost be $C_2HD$; in fact, a larger fraction is lost as $C_2HD$.

Let us continue to examine the reactivity of olefins to make another point. To do this, we need to discuss the reactivity of olefin-radical cations, and so we step away briefly once again from a close analogy to simple ionic chemistry in solution. In the higher olefins, a comparable four-center mechanism

$$R_1CH=CHR_2 \atop +\cdot \atop R_3CH=CHR_4 \longrightarrow \overset{+}{R_1CH-CHR_2} \atop | \atop R_3CH-\overset{\cdot}{C}HR_4$$

accounts for many reactions of ion-molecule collision complexes. The details of such reactions may be easily understood if we briefly review factors affecting the stabilities of carbonium ions.

From solution studies, a tertiary carbonium ion is more stable than a secondary carbonium ion, and that more than a primary carbonium ion.

$$(CH_3)_3C^+ > (CH_3)_2CH^+ > CH_3CH_2^+$$

$$(CH_3)_3C^+ > CH_3\overset{+}{C}HCH_2CH_3 > CH_3CH_2CH_2CH_2^+$$

This ranking is reflected in many properties, rate constants, changes of mechanism, and so forth. It carries into the gas phase with ionization potentials of radicals, and qualitative theories of mass spectral fragmentation, and so the order in solution is a fundamental property of the ions and not a result dominated by solvent interactions. Presumably its origin is in the greater polarizability of the more branched structures.

Thus in the complex for 2-butene-$d_8$ and terminal olefins the intermediate

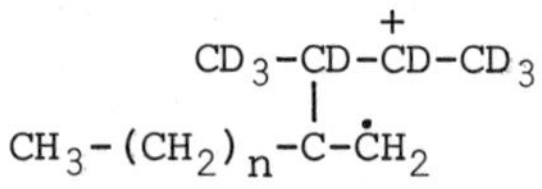

$$CD_3-\overset{+}{CD}-CD-CD_3$$
$$|$$
$$CH_3-(CH_2)_n-C-\overset{.}{C}H_2$$

cannot be divided readily on paper to give a $C_5$ fragment, but
the complex drawn according to this rule

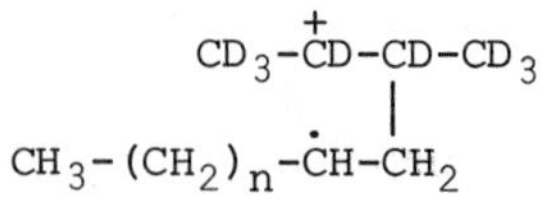

$$CD_3-\overset{+}{C}D-CD-CD_3$$
$$|$$
$$CH_3-(CH_2)_n-\overset{.}{C}H-CH_2$$

can, and it gives in particular $C_5H_2D_8^+$, the major labeled
fragment observed. A number of other similar systems were
tested, and both the isotopic distribution and the ease of loss
of $C_1$, $C_2$, or $C_3$ fragments favored the key intermediate of the
type suggested; in each case polarizability of the alkyl
substituents on the cationic and radical site and steric factors
can be used to justify the selection of the intermediate giving
the major products.

In these cases, however, the lifetime of the intermediate
does not seem to govern the H, D exchange. One would anticipate
that very exothermic, and therefore fast, reactions should show
less exchange than less exothermic, slower processes, because of
more successful competition with an exchange process in the
latter case. This is not found. Exchange seems to be most
facile for hydrogens bound to the carbons bearing the charge and
free radical. In the intermediate formed from 2-butene-$d_8$ and
2-hexene,

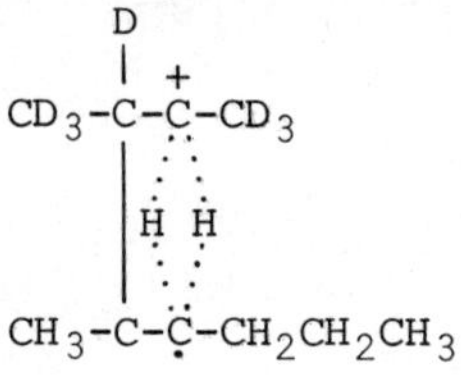

$$\overset{D}{\underset{|}{|}}\overset{+}{\underset{}{}}$$
$$CD_3-C-C-CD_3$$
$$H \; H$$
$$CH_3-C-\overset{.}{C}-CH_2CH_2CH_3$$

the exchange indicated is the only important one needed to
explain product isotope distribution. Thus instead of random
scrambling, a special mechanism reflecting the postulated
reaction complex shown above for addition to olefins is indicated.[E22,E23]

Finally, there are some interesting reactions that seem
analogous to the Diels-Alder reaction in solution, in which a
1,3-diene reacts with certain olefins (dienophiles). For
example, butadiene and ethylene in solution give cyclohexene.

In the ICR cell, the molecular ion of butadiene has been found to react with a number of olefins. Among the $C_5H_{10}$ isomers below, only the first five gave products with more carbons.

These were $C_8H_{13}^+$, $C_7H_{12}^{+\bullet}$, $C_7H_{11}^+$, $C_6H_{10}^{+\bullet}$, and $C_6H_9^+$ respectively. The sole reaction of 2-methyl-2-butene (which has the lowest ionization potential) was charge exchange, and cyclopentane did not react at all. In the case of the first five compounds, differing amounts of product were produced (see Table 3-8).

So the reaction of $C_4H_6^{+\bullet}$ with at least these olefins has a degree of specificity. No reaction occurred with the saturated

Table 3-8. Relative Amounts of Ions Formed at Low Pressure in Mixtures of Butadiene and $C_5H_{10}$ Isomers[a]

| M/e | 1-Pentene | 3-Methyl-1-Butene | 2-Methyl-1-Butene | cis-2-Pentene | trans-2-Pentene |
|---|---|---|---|---|---|
| 81 | 11.4 | 23.9 | 8.4 | 13.0 | 7.4 |
| 82 | 47.5 | 50.7 | 100 | 27.9 | 29.3 |
| 95 | 12.5 | 11.9 | 41.1 | 70.3 | 39.5 |
| 96 | 33.3 | 23.2 | 10.8 | 8.9 | 11.2 |
| 109 | 1.7 | 14.0 | 23.4 | 16.0 | 13.6 |

[a]Data independent of $C_5H_{10}$ pressure up to $10^{-4}$ torr and corrected for interferences.

isomer, nor did anything of great interest occur with the trisubstituted compound. It is unclear whether this latter point is a consequence of a steric interaction or whether the facile charge exchange opens a new reaction channel sufficiently favorable in energetics to independently explain the loss of higher ion-molecule products.[E49]

The terminal olefins give more $C_6$ ions than $C_7$, and the 2-pentenes give more $C_7$ than $C_6$. A simplistic approach assuming retention of structural identity and a Diels-Alder process explains the greater portion of this product formation,

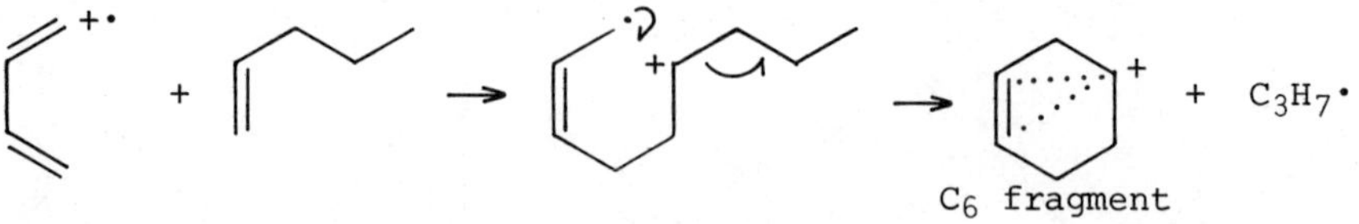
+ $C_3H_7$•
$C_6$ fragment

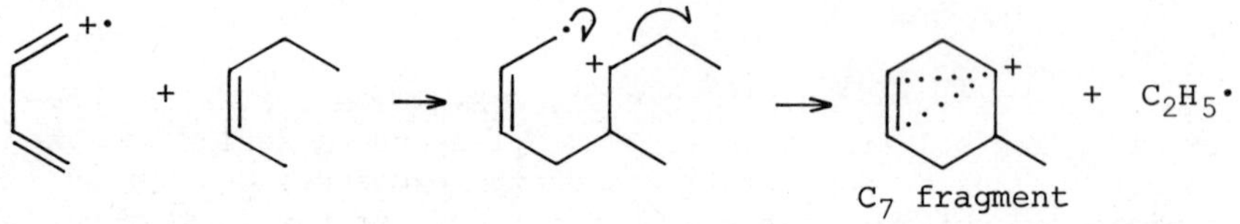
+ $C_2H_5$•
$C_7$ fragment

But not all!

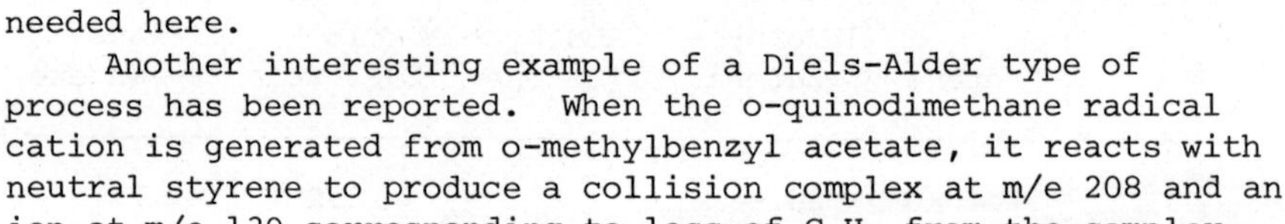
+ $C_2H_5$•
$C_7$ fragment!

The analogy is not perfect. Clearly, further study is needed here.

Another interesting example of a Diels-Alder type of process has been reported. When the o-quinodimethane radical cation is generated from o-methylbenzyl acetate, it reacts with neutral styrene to produce a collision complex at m/e 208 and an ion at m/e 130 corresponding to loss of $C_6H_6$ from the complex.

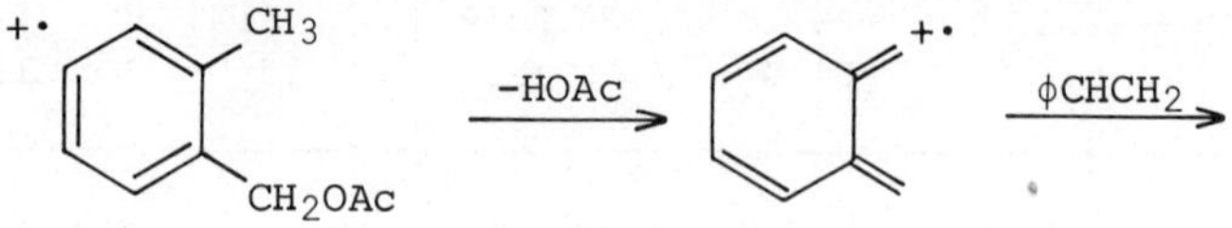
$CH_3$
$CH_2OAc$
-HOAc
$\phi CHCH_2$

$$C_{16}H_{16}^{+\cdot} \xrightarrow{\ -C_6H_6\ } C_{10}H_{10}^{+\cdot}$$
$$\text{m/e 208} \qquad\qquad\qquad \text{m/e 130}$$

A variety of deuterium-labeling experiments involving the ion, styrene, and 2-phenyltetralin, the suspected intermediate, indicate a scrambling reaction primarily between the *ortho* hydrogens of the 2-phenyl group and the two hydrogens in the 4 position of the tetralin. This exchange is followed by competitive 1,2 and 2,4 elimination of $C_6H_6$, the latter pathway dominating.

The reaction occurs in the intermediate and in the conventional mass spectra of the variously labeled 2-phenyltetralins.[20]

The reaction is formally similar to a previously described reaction[D14] of styrene molecular ion with neutral styrene; the masses of the products are similar. However the o-quinodimethane ion is about 15 times more reactive than styrene ion, and the latter ion gives a product from which $C_6H_6$ is lost with essentially no scrambling by a 1,4-mechanism. The product is similar to 1-phenyltetralin in its reactivity.

*Esterification*

The analogy in ICR to solution esterification, and acylation in general, has not yet been worked out in detail.  There are several processes known in acid solution.  A common process is:

$$CH_3CO_2H + H^+ \longrightarrow CH_3-CO_2H_2^+$$

Another process for certain sterically hindered acids is:

Intermediates similar to some of those generated in

solution have been found in the ICR cell, and their reactivity
resembles that of the corresponding solution species.   In
mixtures of small acids and alcohols, the following processes
are observed in the gas phase.  For acid alone in the cell:

$$CH_3COOH^{+\cdot} + CH_3COOH \longrightarrow CH_3CO_2H_2^+ + CH_2CO_2\cdot$$

When an alcohol is added to the acid in the cell, products of
the following reaction are found to come from protonated acetic
acid.  Applying the polarizability rule found earlier (pp. 59

$$CH_3-C\diagup\!\!\!\!\!\overset{O}{\diagdown}_{OH_2^+} + CH_3OH \longrightarrow \left[ CH_3-C\overset{OH}{\underset{CH_3OH}{\diagdown OH}} \right]^+ \longrightarrow CH_3-C\diagup\!\!\!\!\!\overset{O}{\diagdown}_{\underset{H}{\overset{+}{OR}}} + H_2O$$

and 60), we expect that the ionic product is more stable than
the ionic reactant.  (The site of protonation is not defined in
the final ionic product.)  The reaction is, if only for thermodynamic purposes, a displacement of $H_2O$ from $CH_3CO^+$ by $CH_3OH$,
and could be used to establish the order of the acetyl ion
affinities of the two neutral molecules.[E66]
     The other process, direct acylation by an acyl ion, is
observed in relatively few cases at low pressure.  These cases
have been found among much work done with acetylation of
oxygen- and nitrogen-containing compounds by ions derived from
biacetyl.[E20,E64]

$$CH_3CO^+ + (CH_3)_2CHCl \longrightarrow (CH_3)_2CHClCOCH_3^+$$

$$CH_3COCOCH_3^{+\cdot} + \underset{N}{\bigcirc} \longrightarrow \underset{\underset{COCH_3}{\overset{+}{N}}}{\bigcirc} + \cdot COCH_3$$

This latter product is similar to the acetylpyridinium ion postulated when pyridine is added as a catalyst to esterification
to a mixture of an alcohol and an acid chloride or anhydride.

$$\overset{\overset{O}{\|}}{CH_3C-O^+}\underset{CH_3-C-C\diagdown CH_3}{\overset{\|}{\phantom{x}}}\diagup\!\!\!\overset{O}{} + C_2H_5OH \longrightarrow C_2H_5O^+\!\diagup\!\!^H \underset{CH_3}{\overset{}{C}}\diagup\!\!\!\overset{O}{} + CH_3COCOCH_3$$

If one wishes to consider the last reactant ion as a solvated
acetyl ion, then there are numerous examples of esterification
by transfer of gaseous acetyl ion from "solvent" to an alcohol.
For the epimeric alcohols

the former (the *exo* compound) was acetylated according to the
last process more rapidly than the latter (*endo*) compound.  This
could be explained by the greater interference by *endo*-hydrogen
toward approach of the reactant ion, or by other steric arguments.[E33]
   In basic solution, a process known as ester interchange

$$RO^- + CH_3\overset{O}{\underset{|}{C}}-OR' \;\rightarrow\; CH_3\overset{O^-}{\underset{\underset{OR}{|}}{C}}-OR' \;\rightarrow\; CH_3\overset{O}{\underset{}{C}}-OR + OR'^-$$

occurs.  The analogy exists in the gas phase.  Negative-ion
studies of ester interchange have been observed in the ICR cell.
Displacement of alkoxide in formate esters occurs if the anion
produced is no less stable than the reagent anion.  For example:

$$CD_3O^- + HC\overset{O}{\underset{}{}}-OCH_3 \;\rightarrow\; H-C\overset{O}{\underset{}{}}-OCD_3 + {}^-OCH_3$$

$$CH_3O^- + HC\overset{O}{\underset{}{}}-OC_6H_5 \;\rightarrow\; H-C\overset{O}{\underset{}{}}-OCH_3 + {}^-OC_6H_5$$

$$C_2H_5O^- + HC\overset{O}{\underset{}{}}-OC_3H_7 \;\rightarrow\; H-C\overset{O}{\underset{}{}}-OC_2H_5 + {}^-OC_3H_7$$

In each case there is also a competing reaction, and it is the
only one in cases where the anion expected as a product is less
stable (less polarizable) than the reactant anion.  Hence
because $^-OCH_3$ is less stable than $^-OC_2H_5$ or $^-OC_3H_7$, we find only
these reactions:

$$C_2H_5O^- + \overset{\overset{\displaystyle O}{\|}}{H}COCH_3 \longrightarrow (C_2H_5O-H-OCH_3)^- + CO$$

$$C_3H_7O^- + \overset{\overset{\displaystyle O}{\|}}{H}C-OCD_3 \longrightarrow (CD_3-O-H-OC_3H_7)^- + CO$$

Yet even here there is an analogy to solution chemistry. Note that this corresponds to decarbonylation of formate esters in strong base![E66]

$$OR^- + \overset{\overset{\displaystyle O}{\|}}{H}C-OCH_3 \longrightarrow ROH + CO + {}^-OCH_3$$

We can conclude by pointing out that many of the "typical" reactions of organic functional groups that proceed through ionic intermediates in solution have been shown by this kind of work to have clear, though limited, analogies in the gas phase. Solvation appears to play a major role in determining the breakdown of the intermediate, since the stabilities of unsolvated species often alter the nature of the products from the well-known products of solution chemistry. Both the close analogies found and the differences are of considerable interest at this stage, because the degree of correspondence is still being delineated.


OTHER ASPECTS OF REACTIVITY


*Ligand Substitution Processes*

The applicability of ICR to transition metal chemistry is indicated by a single study involving displacements on iron pentacarbonyl by various ligands.
  Mixtures of $Fe(CO)_5$ and various ligands, HCl (PA 140 kcal/mole), $CH_3F$ (PA 151), $H_2O$ (PA 164), and $NH_3$ (PA 207), have been examined with an eye to establishing the kinds of reaction that ions derived from these compounds undergo. The relative rates of ligand substitution upon different positively charged ions $Fe(CO)_n^+$ are given in Table 3-9.
  Hydrogen chloride fails to react. For the others, the metal-ligand binding energy is apparently greater than that of Fe-CO, so that reaction occurs. The rates are not linearly related to the metal-ligand binding energies, however. If one

Table 3-9.   Relative Rates of the Process
$L + Fe(CO)_n^+ \rightarrow Fe(CO)_{n-1}L^+ + CO$

| L | n = 1 | n = 2 | n = 3 | n = 4 | n = 5 |
|---|---|---|---|---|---|
| HCl | 0 | 0 | 0 | 0 | 0 |
| $CH_3F$ | 0.45 | 0.82 | 1 | 0.16 | <0.01 |
| $H_2O$ | 0.86 | 0.91 | 1 | 0.56 | <0.01 |
| $NH_3$ | 0.92 | 0.98 | 1 | 0.56 | 0.16 |

compares rates with the proton affinities given above, there does appear to be a correlation with proton affinity, but it is only qualitative.

Fe(CO)$_4^+$ also reacts with *benzene*, in an interesting process.

$$Fe(CO)_4^+ + C_6H_6 \rightarrow Fe(CO)_2(C_6H_6)^+ + 2CO$$

Negative ions displace carbon monoxide from Fe(CO)$_5$. Both $F^-$ and $C_2H_5O^-$ produce XFe(CO)$_3^-$ with release of two CO molecules.

Condensation of Fe(CO)$_5^+$ with Fe(CO)$_5$ to give Fe$_2$(CO)$_9^+$ + CO has been noted as the first gaseous ionic formation of a binuclear complex.

Among the reactions of charged small molecules with neutral Fe(CO)$_5$, $(CH_3)_2F^+$ produces CH$_3$Fe(CO)$_5^+$ and CH$_3$Fe(CO)$_4^+$, the former species showing an increased oxidation state of Fe. This is then a gaseous redox reaction of a transition metal!

The H$_3$O$^+$ ion transfers a proton, giving HFe(CO)$_5^+$ and HFe(CO)$_4^+$; these also are formed from H$_2$Cl$^+$. A proton is not transferred from ammonia. This places rather wide proton-affinity limits on Fe(CO)$_5$:   207 kcal > PA$\left(Fe(CO)_5\right)$ > 164 kcal.[E30] The limits for ferrocene have been narrowly defined and the surprisingly high (similar to methylamine) proton affinity of 217 kcal assigned.[21] The poor solvation of this species accounts for the wide variation between its solution basicity and that of methylamine.

This brief study, in which ligand displacement, a redox process, and formation of a binuclear complex were reported, suggests a fruitful area of future research.

*A Negative Ion-Molecule Reaction in Diborane*

The detection of interesting intermediates in boron chemistry
may help to elucidate the solution chemistry of these compounds.
In addition to products formed in low amounts by pyrolysis at
the filament, a scan of diborane anions shows $BH_4^-$ and $B_2H_7^-$.
Double-resonance studies on the latter ion indicate that it is
formed from $BH_4^-$, so that a reaction may be written

$$BH_4^- + B_2H_6 \longrightarrow B_2H_7^- + BH_3$$

The intermediate complex in this case is $B_3H_{10}^-$. Because boron
has two isotopes, $^{10}B$ and $^{11}B$, an analysis of the double-
resonance signals may be made. A matrix analysis based upon the
method used for the alkyl-chloride reaction discussed in the
previous section suggests that all three boron atoms in this
complex are equivalent. That is, the $BH_3$ neutral lost from the
complex is equally likely to contain any of the boron atoms of
the reactants.
    The species suggested for this intermediate is not
symmetric. Instead, rapidly equilibrating species are
indicated, these being more in keeping with possible analogies
in bonding in boron hydrides. They are:

The tautomerism must be rapid, since $B_3H_{10}^-$ is not observed in
the spectrum. This absence places the upper limit of its lifetime at $10^{-4}$ s.
    The addition of water to the diborane does not produce
oxygen-containing species but rather $B_2H_5^-$, which apparently
results as a secondary process. The $OH^-$ ion apparently produces
$BH_4^-$ with considerable internal excitation, and leads to the
second reaction

$$OH^- + B_2H_6 \longrightarrow BH_4^{-*} + \text{neutral}$$

$$BH_4^{-*} + B_2H_6 \longrightarrow B_2H_5^- + BH_3 + H_2$$

Thus the formation of $BH_4^{-*}$ from $OH^-$ with excess internal energy
opens a new route of higher energy. This point--that the
channels through which a complex decomposes depend on its

internal energy--is discussed in Chapter 4.[E5]

*Structures of Gaseous Ions: Symmetry of Some Small Ions and
Intermediates*

The techniques available for the study of the structure of
gaseous ions are limited, and for this reason alone such
problems are intriguing.  In the case of relatively common
species found in the mass spectrometer, the problem is also of
considerable importance, not only because a thorough understanding will allow clearer interpretation of analytically useful
reactions, but also because it will assist in the interpretation
of processes caused by other kinds of radiation.
          Consider first the structure of the small hydrocarbon ion
$C_3H_6^{+\cdot}$.  The ion formed from cyclopropane reacts with neutral
ammonia[D5,D12]:

$$C_3H_6^{+\cdot} + NH_3 \quad \nearrow \quad CH_5N^{+\cdot} + C_2H_4$$
$$\searrow \quad CH_4N^+ + C_2H_5\cdot$$

but that formed from propylene does not:

$$C_3H_6^{+\cdot} + NH_3 \longrightarrow \text{ no reaction}$$

This implies that they have different structures or very
different internal energy content.  The latter hypothesis was
made less likely when it was found that the reactive state could
be reached from a variety of precursors throughout a broad range
of ionizing energies.
          The structure of the reactive $C_3H_6^{+\cdot}$ is not necessarily
that of cyclopropane; the only information is that it is most
likely not the same as that of $C_3H_6^{+\cdot}$ from propylene, whatever
that may be.  In order to learn more about its structure,
labeling experiments were performed.
          In the first place, it was shown that there is no
scrambling of H atoms in the collision complex between $C_3H_6^{+\cdot}$
and $NH_3$; the reactions

$$C_3H_6^{+\cdot} + ND_3 \quad \nearrow \quad CH_2D_3N^{+\cdot} + C_2H_4$$
$$\searrow \quad CH_2D_2N^+ + C_2H_4D\cdot$$

are nearly exclusive.
          With such a clean system, it is possible to label the
$C_3H_6^{+\cdot}$ ion in order to interpret results quickly, but of course

one cannot find a carbon distinguishable by its reactivity in cyclopropane to label.  They are all identical, and if the structure of reactive $C_3H_6^{+\cdot}$ is $\cdot CH_2CH_2CH_2^+$, where they are not identical, one cannot start from cyclopropane-1,1-$d_2$ to make a specifically labeled ion.  The problem of specific labeling was answered by taking advantage of the observation that the fragment ion from tetrahydrofuran behaves identically to the molecular ion of cyclopropane over wide energy and pressure

$$\left[ \begin{array}{c} \text{⬠} \end{array} \right]^{+\cdot} \xrightarrow{\ -CH_2O\ } \ C_3H_6^{+\cdot}$$

ranges, and therefore must have the same structure.  Suppose this structure is linear.  It would be expected to react with ammonia in the following way:

$$\left[ \begin{array}{c} \text{D}\diagdown\diagup\diagdown\text{H} \\ \text{D}\diagdown_{\text{O}}\diagup\diagdown\text{H} \end{array} \right]^{+\cdot} \rightarrow \ \begin{array}{c} \text{CH}_2\cdot \\ \text{D}\diagdown\diagup\diagdown \\ \text{D}\diagup\diagdown_{\overset{+}{\text{O}}}\diagup^{\text{CH}_2} \end{array} \longrightarrow \ \cdot CH_2CH_2CD_2^+ \xrightarrow{\ NH_3\ }$$

$$\cdot CH_2CH_2CD_2NH_3^+ \begin{array}{c} \longrightarrow \ CD_2NH_3^{+\cdot} + CH_2CH_2 \\ \longrightarrow \ CD_2H_2N^+ + C_2H_5\cdot \end{array}$$

and the pathways can also be checked using $ND_3$ as the reagent gas.

Yet in fact nearly statistical scrambling of the hydrogens originally in $C_3H_6^{+\cdot}$ was observed.  This implies that $C_3H_6^{+\cdot}$ must be in the cyclic form, if not when it reacts, at least at some point in its history in passing from one acyclic form to another, so that the $CH_2$ groups are randomized.

$$\cdot CH_2CH_2CD_2^+ \ \rightleftharpoons \ \left[ \begin{array}{c} \text{CD}_2 \\ \diagup\diagdown \\ \text{CH}_2\!-\!\text{CH}_2 \end{array} \right]^{+\cdot} \ \rightleftharpoons \ \cdot CH_2CD_2CH_2^+, \ \text{etc.}$$

The conclusion is that the cyclic structure is sufficiently stable to be attained by all $C_3H_6^{+\cdot}$ ions from cyclopropane and tetrahydrofuran that undergo these reactions with ammonia. (That statement does not necessarily imply that the ions which do *not* react have the same structure!)

We also should note that this kind of technique is useful

for demonstrating the nature of collision complexes as well as
product ions.  Here is an early example, the reaction of an
olefin with hydrogen sulfide.  Mixtures of ethylene and hydrogen
sulfide produce a peak at m/e 47, $H_2CSH^+$, whose precursors are
$C_2H_4^{+\cdot}$ and $H_2S^{+\cdot}$.  With labeled reagents, experiments typified by
the following scheme established the nature of the reaction:

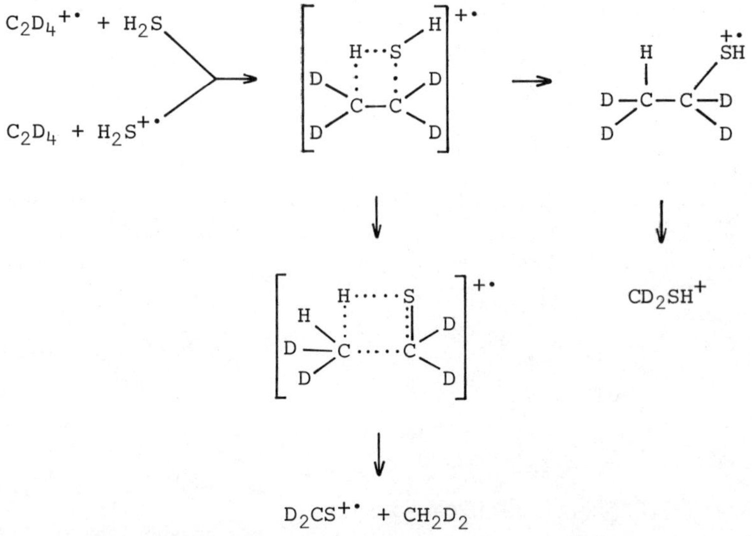

This is a mixture of processes, and must be influenced by the
radical nature of one reactant in any case, irrespective of the
order of addition of charged species.  Some scrambling of
hydrogen and deuterium among carbon and sulfur atoms occurs, but
these processes do not obscure the fundamental chemistry of the
addition reaction illustrated above.[E15]

Addition to triple bonds is a simple extension of the
previous case.  The reaction of acetylene and hydrogen sulfide
in the ICR cell is analogous in many respects:

$$\left[\begin{array}{c} H-S-H \\ + \\ D-C\equiv C-D \end{array}\right]^{+\cdot} \longrightarrow \left[\begin{array}{c} H \\ H \diagdown \quad S \\ \quad C=C \\ D \diagup \quad \diagdown D \end{array}\right]^{+\cdot} \longrightarrow \left[\begin{array}{c} H \\ | \\ S \\ H \diagdown \quad | \\ \quad C=C \\ D \diagup \quad \diagdown D \end{array}\right]^{+\cdot}$$

$$\left[\begin{array}{c} H \\ H—C-C \diagup^{S} \\ D \quad \diagdown D \end{array}\right]^{+\cdot} \qquad \begin{array}{c} H \\ | \\ S \\ \diagup \diagdown \\ C———C^{+\cdot} \\ D \diagup \qquad \diagdown D \end{array}$$

$$\downarrow \qquad\qquad\qquad \downarrow \text{ -H or D}$$

$$DCS^{+} + CH_2D\cdot \qquad\qquad \begin{array}{c} H \\ | \\ S^{+\cdot} \\ \diagup \diagdown \\ C===C \\ R \diagup \qquad \diagdown H \end{array}$$

Again, there is some scrambling of hydrogens in the thio-acetaldehyde ion, but the cleavage to form thioformyl ion ($DCS^{+}$) is at least six times faster, so that the major ionic product is $DCS^{+}$, not $HCS^{+}$. Therefore the structure of the complex just before cleavage is mostly that of thioacetaldehyde. The addition to give a thioacetaldehyde ion eventually requires a rearrangement, but it is at least loosely analogous to the addition of water to acetylenes catalyzed by mercuric sulfate:

$$RC\equiv CH \xrightarrow[\text{HgSO}_4]{\text{H}_2\text{O, H}_2\text{SO}_4} R-\overset{\overset{\displaystyle O}{\|}}{C}-CH_3$$

The analogous processes in mixtures of ethylene or acetylene and water have not been reported, however; only some of them are calculated to be exothermic.

*Structures of Some Gaseous Ions: The McLafferty Product Ion*

As a further example of structural analysis, we choose products of the McLafferty rearrangement, which is the most common rearrangement producing useful information in mass spectrometry.[22]

A good example is the reaction of the 2-hexanone molecular ion.

In addition, the consecutive operation of two such processes is known in molecules of appropriate structure

We cannot present all lines of evidence used to explore this reaction, but the experiments recently performed by ICR assisted considerably in elucidating structures of the product ions.

There are several reactions in which the McLafferty product ion is represented as the enol of acetone and differs in further reactivity from the species produced by the direct ionization of acetone. A mixture of acetone and another ketone that undergoes the McLafferty reaction was used in the ICR experiments. For structures A and B, simultaneously present in the ICR cell,

A                                    B

A  Acetylates acetone and transfers charge to 2-hexanone.
B  Transfers a proton to 2-hexanone.

A  Acetylates the 5-nonanone, with loss of 2 $C_3H_6$.
B  Protonates 5-nonanone.

Only B condenses, with loss of water and either methyl radical or ethylene, with 1-methylcyclobutanol.

The first set of these reactions is analogous to the reactivity of ketones and alcohols respectively upon ionization,

$$\overset{O^{+\cdot}}{\underset{\|}{R-C-R}} \;+\; \overset{O}{\underset{\|}{R-C-R}} \;\longrightarrow\; \overset{+OCOR}{\underset{\|}{RCR}} \;+\; R\cdot$$

$$ROH^{+\cdot} \;+\; \overset{O}{\underset{\|}{R-C-R}} \;\longrightarrow\; RO\cdot \;+\; \overset{HO^{+}}{\underset{\|}{RCR}}$$

and constitutes evidence for a difference in structure (different chemistry implies different compounds unless there are large internal-energy effects) and good evidence that the structures are as written. Additionally, the ion formed by loss of ethylene from the molecular ion of 1-methylcyclobutanol

$$\underset{CH_2-CH_2}{\overset{+\cdot}{\underset{\text{OH}}{CH_2-C}}} \overset{}{\underset{CH_3}{}} \;\longrightarrow\; CH_2=C\overset{+\text{OH}}{\underset{CH_3}{}} \;+\; CH_2=CH_2$$

undergoes exactly the same reaction as B; this suggests that they have the same structure. Finally, the ion formed from two consecutive McLafferty reactions (two losses of $C_3H_6$ in 5-nonanone) also undergoes exactly the same reactions as B, and so should have the same structure, which we illustrated at the beginning of this section. These ions were taken to have identical structures as a result.

The last case was of interest because there had been some debate among mass spectrometrists about the structure of the product ion after two losses of $C_3H_6$. It had been suggested that an oxonium ion was a possible structure.

$$\underset{H}{\overset{H}{>}}\overset{+}{O}-C\overset{CH_2}{\underset{\cdot CH_2}{<}} \;\longleftrightarrow\; \underset{H}{\overset{H}{>}}\overset{+}{O}-C\overset{\cdot CH_2}{\underset{CH_2}{<}}$$

This latter possibility was discredited by the following reactions. In any mass spectrometer or ICR spectrometer at sufficiently high ionizing energy, 4-nonanone, in keeping with the known reactivity of dialkyl ketones, loses the larger alkene because secondary hydrogen transfer occurs 10-20 times as often as primary hydrogen transfer.

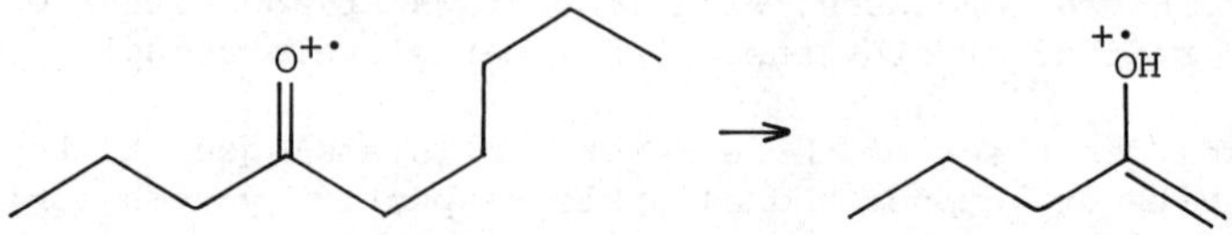

This ion then loses ethylene to give the ion of interest $(C_3H_6O^+)$.

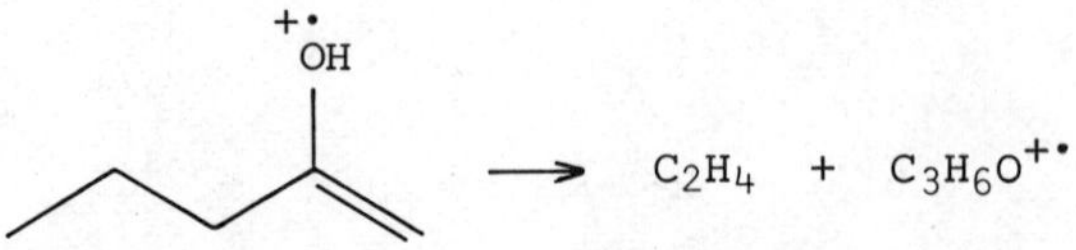

$$C_2H_4 \;+\; C_3H_6O^{+\bullet}$$

Suppose, however, that we use 4-nonanone-1,1,1-$d_3$. Then our first product can give the following $C_3H_6O^{+\bullet}$ ions depending on the mechanism of the last step.

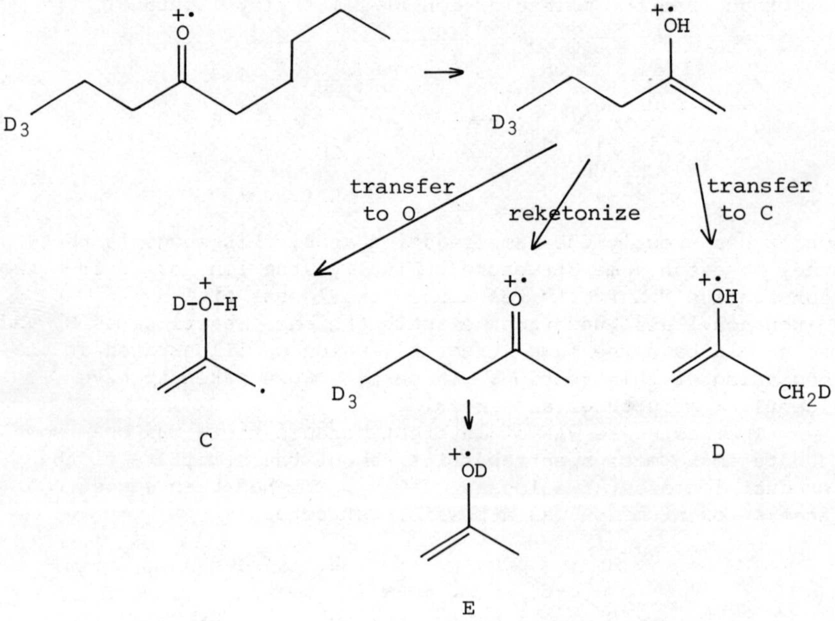

transfer to O          reketonize          transfer to C

C

D

E

Only enolic hydrogen is transferred easily to other species, as extensive labeling work with the single McLafferty reactions described above had shown for the molecular ion of acetone and its enol.  Hence C will transfer nearly equal amounts of H and D

$$CH_2\!=\!C(OD)CH_3^{+\cdot} + RCOR \longrightarrow RCODR^+ + CH_2\!=\!C(CH_3)O\cdot$$

(subject to an isotope effect), D will transfer only H, and E will transfer only D.  In fact, hydrogen transfer to neutral 4-nonanone and other ketones is favored by a 5:1 margin.  On the assumption that the special ability of enolic H to transfer still predominates here, the result suggests either that the pathway involving D is the most favored, or that a very large isotope effect exists.

   The latter point was shown not to be important by studying the same reaction in 4-nonanone-7,7-$d_2$.  In this case the expected transfer of hydrogen is as follows.  The oxonium ion C' transfers either H or D, the enol D' transfers D, and the enol E' transfers H.  That is,

Deuterium was transferred six times as readily as hydrogen.
Thus the isotope effect is unimportant in this case, and D (D')
is the most appropriate representation.

Similar confirmation of these results was obtained by
examining the condensation with l-methylcyclobutanol in which
ethylene and water are lost.  In this reaction, labeling
experiments had shown the loss of enolic hydrogen.  The results
were consistent with the above interpretation.

After that last confirmation, let us back up and explore a
fine point of the previous paragraph.  Why a 6:1 ratio of D/H
transfer from D'?  Why not 100% D transfer?  The reaction that
produces the transfer of the minor isotope was suggested to be a
result of a small amount of hydrogen scrambling and/or the
production of ions from the *less* favored first step, in which
*primary* hydrogen is transferred to O.

Scrambling could be more important at low ionizing voltage.
Thus the oxonium structure can contribute only a minor amount to
the population of $C_3H_6O^{+\cdot}$.  There are three unlikely loopholes
in this argument, but they are corequisite; the oxonium ion:
(a) may not transfer protons to the bases used, (b) does not
undergo the condensation with l-methylcyclobutanol, and (c) does
not transfer charge to 2-hexanone.[D1,D2,D3]

At least one ketone does reketonize.  The ion formed from
2-propylcyclopentanone undergoes the proton-transfer reaction

indicative of enolic structure at short residence times
($1 \times 10^{-4}$ s) with several neutral ketones. These reactions,
which were followed by double resonance in the analyzer,
disappear when the lifetime in a conventional cell is increased
to $10^{-1}$ s, and are replaced by reactions (in the analyzer) that
are more suggestive of ketonic behavior.

This suggests that increasing the residence time in the source
depletes the enolic content of $C_5H_8O^{+\cdot}$ by a process that con-
verts enol ions to a ketonic form.

The reactivity of the enol from 2-hexanone, however, is not
changed by increasing the residence time up to $10^{-1}$ s.[E42]

*Photochemistry of Some Gaseous Ions*

The ICR spectrometer offers a unique opportunity to study the photochemistry of gaseous ions. Because ions can be trapped for as long as several seconds in the cell, it is possible to hold ions in a photon beam after they have been formed, and, if the energy of the incident light is sufficient, to initiate chemical reactions. One can, as a corollary, vary the wavelength of the incident light in order to accurately measure the energy thresholds of endothermic reactions.

Interesting experimental results for small molecules have been included in Chapter 5, where they can be discussed in the language of chemical physics more easily than is the case of the larger molecules of greater interest to the organic chemist.

Some very interesting studies of the interaction of light with hydrocarbons have been carried out. These hydrocarbons were chosen because the structures of ions derived from them have vexed organic mass spectrometrists since the earliest physical-organic studies in that field. In particular, the chemistry of toluene on ionization by electron impact is not trivial. Before it undergoes its least-energy dissociation, loss of H, the toluene cation rearranges to a structure from which any hydrogen atom is lost with the same probability. This problem has been elegantly explored by conventional mass

spectrometry and related techniques, and we omit some detail in
summarizing the present concept of the actual process.

It is possible to ionize toluene in the ICR cell with
electrons of such low energy that $C_7H_8^{+\cdot}$ ions are formed with no
more than a small amount of vibrational energy. Irradiation of
toluene cations formed in this way produces a loss of hydrogen,

$$C_7H_8^{+\cdot} \xrightarrow{h\nu} C_7H_7^+ + H\cdot$$

and the ion formed reacts further:

$$C_7H_7^+ + C_7H_8 \longrightarrow C_8H_9^+ + C_6H_6$$

The onset for this dissociation is at 5200 Å, no maximum being
found below 4000 Å. The rate of the reaction with neutral
toluene also increases steadily with energies corresponding to
the range 5000 to 4000 Å. The onset value for observation of
$C_7H_7^+$ corresponds to an electronically excited state of the
cation about 2.2 eV above the ground state, which is also
observed in the photoelectron spectrum of toluene. After onset,
there is a very sharp rise in the $C_7H_7^+$ signal, which agrees
with theory because both the ground and excited states of $C_7H_8^{+\cdot}$
correspond to removal of a pi electron. Their geometries would

then be similar, and Franck-Condon factors for photons near
threshold ought to be favorable.

Labeling experiments (using $C_6H_5CD_3$ and $C_6D_5CH_3$) indicate
the loss of hydrogen at the same rate irrespective of its
original position in the toluene molecule at any wavelength
between 4000 and 5200 Å. There is, however, an isotope effect
that increases with decreasing cation energy, in reasonable
agreement with theoretical predictions and with the isotope
effect observed for dissociation under electron-impact conditions in a mass spectrometer.[E51]

These results are in agreement with the results from
classical mass spectrometry. If sufficient energy is supplied
to $C_7H_8^{+\cdot}$ to dissociate it, there is also sufficient energy to
isomerize the structure so that the hydrogens are scrambled. Is
it possible that hydrogens scramble at all energies? This
question cannot be answered by conventional mass spectrometry,
because there we can examine molecules only by their dissociation. Molecules that do not dissociate can be examined by ion
molecule reactions in the ICR spectrometer.

The mass spectra of norbornadiene (A) and cycloheptatriene
(B) are nearly the same as that of toluene, of which they are

A                                    B

isomers. This has prompted the suggestion, reinforced by
appropriate experiments, that the ions formed from them by
removal of an electron attain a common set of structures before
they decompose. On the other hand, it has been shown by a
conventional ICR experiment that a fraction of the toluene ions
do not attain the same structure as cycloheptatriene or
norbornadiene.[D7] Toluene molecular ion reacts with isopropyl
nitrate while the other molecular ions do not. In support of

this, the photodissociation curves of the three are entirely
different in the range between 4000 and 8000 Å. Hence in this
range of energies, the three species do not interconvert to any

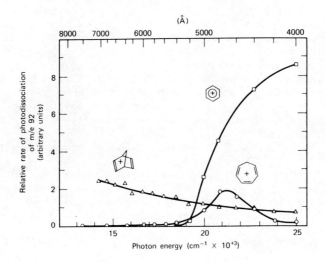

Figure 3-4.   Relative photodissociation rates of $C_7H_8^{+\cdot}$ ions as a function of wavelength for toluene (□), cycloheptatriene (o), and norbornadiene (Δ).  The three curves are normalized to an arbitrary common vertical scale and may be compared directly. (Reprinted, by permission, from *J. Amer. Chem. Soc. 95*, 2716 (1973), Figure 1.)

appreciable extent because their chemistries differ.  Interconversion must therefore have a relatively high threshold[E53] (see Figure 3-4).

    To summarize, these experiments have clearly illustrated the detailed information about low-energy ions that can be derived from photodissociation studies.  In a sense, the experiments are a sort of recording of absorption spectra, where detection is by monitoring a chemical reaction product.  This is exciting work.

## NOTES

1. The excellent work from P. Kebarle's laboratory is particularly noteworthy for its application to organic chemistry.
2. J. L. Franklin, J. G. Dillard, H. M. Rosenstock, J. T. Herron, K. Draxl, and F. H. Field, Ionization potentials, appearance potentials, and heats of formation of gaseous positive cations, U.S. Government Printing Office, Washington, D.C. (1969).
3. A reaction that would be slightly endothermic from the ground state of an ion can be observed when the ionic species has excess internal energy. Such a case could arise, for example, when the ionic species is formed by the fragmentation of a larger ion.
4. R. J. Cotter and W. S. Koski, *J. Chem. Phys.* 59, 784 (1973).
5. W. J. Hehre and J. A. Pople, *J. Amer. Chem. Soc.* 92, 2191 (1970).
6. R. J. W. Le Fèvre, *Adv. Phys. Org. Chem.* 3, 1 (1965).
7. J. P. Briggs, R. Yamdagni, and P. Kebarle, *J. Amer. Chem. Soc.* 94, 5128 (1972).
8. R. Yamdagni and P. Kebarle, *J. Amer. Chem. Soc.* 95, 4050 (1973).
9. R. W. Taft, personal communication.
10. R. Yamdagni, T. B. McMahon, and P. Kebarle, *J. Amer. Chem. Soc.* 96, 4035 (1974).
11. G. W. A. Milne and M. J. Lacey, *Crit. Rev. Anal. Chem.* 4, 45 (1974).
12. With respect to structures of cations, labeling experiments have shown that only the (initially) methyl protons of ethyl cation generated from $(C_2H_5)_2NNO$ are transferred to bases.[D9] This excludes equilibration through a symmetrical protonated ethylene structure for $C_2H_5^+$, which had been considered seriously by theoreticians:

$$CH_3CD_2^+ + B \longrightarrow HB + CH_2CD_2$$

13. J. R. Hass, Ph.D. dissertation, University of North
    Carolina, Chapel Hill (1972).
14. J. L. Beauchamp, Ph.D. dissertation, Harvard University
    (1968).
15. R. Yamdagni and P. Kebarle, *J. Amer. Chem. Soc. 95*, 3504
    (1973).
16. T. B. McMahon, R. J. Blint, D. P. Ridge, and J. L.
    Beauchamp, *J. Amer. Chem. Soc. 94*, 8934 (1972).
17. L. B. Young, E. Lee-Ruff, and D. K. Bohme, *J. Chem. Soc.
    Chem. Commun.*, 35 (1973).
18. F. Bernardi and W. J. Hehre, *J. Amer. Chem. Soc. 95*, 3078
    (1973).
19. D. A. Chatfield, University of North Carolina, unpublished
    observations.
20. M. L. Gross, private communication.
21. J. L. Beauchamp, private communication.
22. D. G. I. Kingston, J. T. Bursey, and M. M. Bursey, *Chem.
    Rev. 74*, 215 (1974).

## CHAPTER 4

Theoretical Aspects of
Ion-Molecule Reactions
and Ion Cyclotron
Resonance Signals


THE RESONANCE PHENOMENON

The term "resonance" appears in a wide variety of contexts in
physics. Among these are: (a) nuclear resonance, in which the
yield of an induced nuclear reaction shows a sharp maximum when
the incident particle has a certain energy, (b) electrical
resonance, in which a circuit shows a sharp minimum in its
resistance to the flow of current of a certain frequency, (c)
acoustical resonance, in which the vibration of an elastic
medium is greatly augmented by a force of a certain frequency,
(d) electron paramagnetic resonance, in which unpaired electrons
absorb energy in the microwave region when placed in a magnetic
field at an intensity of 10 000 gauss.

The list could be considerably lengthened. Three others of
great importance are: (a) NMR spectroscopy, (b) Mössbauer spectroscopy, and (c) light absorption in infrared, visible, and UV
spectroscopy.

The common factor in all of these is the presence of an
energy source and an oscillating system that absorbs energy
subject to a certain condition. At this resonance condition the
system is in forced harmonic oscillation, and the energy source
is said to drive the oscillator.

We can write down the mathematics appropriate to resonating
systems by first dealing with whatever can interrupt the resonance. This interruption is called "damping," and is not
difficult to treat for the special case in which it is proportional to the velocity of the resonating system. The equation
for the forces is:

$$m \frac{d^2x}{dt^2} + c \frac{dx}{dt} + kx = F \tag{1}$$

The first term on the left is the acceleration, the second is the damping, and the third is the Hooke's-law term characteristic of a harmonic oscillator. Their sum equals the driving force F. Division of Equation (1) by m puts it into standard form:

$$\frac{d^2x}{dt^2} + \frac{c}{m}\frac{dx}{dt} + \frac{kx}{m} = \frac{F}{m} \tag{2}$$

For convenience we set $\gamma = c/m$ and $\omega_0^2 = k/m$, where $\omega_0$ is the natural angular frequency of a harmonic oscillator in rad/s. It comes from the solution of Equations (1) or (2) when c = F = 0. The units of c/m are $s^{-1}$.

If, as is frequently the case, the force F is given by a constant multiplied by a cosine function of the time, the energy absorbed by the oscillating system will depend on the angular frequency of F according to a function of the form:[1]

$$\rho^2 = \frac{1}{m^2\left((\omega^2 - \omega_0^2)^2 + \gamma^2\omega^2\right)} \tag{3}$$

Additional constants in Equation (3) adapt it to the particular type of resonance. A graph of this function appears in Figure 4-1. When $\gamma$ is small, it is the width of the curve at half-height.

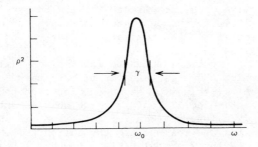

Figure 4-1.   Resonance energy as a function of frequency.

We recall that $\gamma = c/m$, and that $c$ is the damping co-
efficient. Mathematically it sets the width of the energy-
absorption curve. Instrumental factors aside, whatever
interrupts the resonant absorption of energy sets the width of
the absorption curve. Pressure (collision) broadening of the
lines observed in atomic emission spectra and the relaxation
times of NMR spectroscopy are well-known examples. In ICR the
damping of energy or power absorption is caused by collisions of
the resonant ion with molecules.

SIGNAL LINEWIDTHS AND ION-MOLECULE COLLISION FREQUENCIES

The basic equation for the instantaneous power absorbed by an
ion in resonance was derived in Chapter 1 for the special case
in which the angular frequency $\omega_1$ of the oscillator was equal to
the angular frequency $\omega_c$ of the ion in cyclotron motion. A more
general equation gives the instantaneous power absorption per
ion at any value of the oscillator frequency[A35]:

$$A(\omega_1) = \frac{\varepsilon_{rf}^2 e^2}{4m(\omega_1 - \omega_c)^2}\left(1 - \cos(\omega_1 - \omega_c)\tau\right) \tag{4}$$

The value of $A(\omega_1)$ is completely negligible unless $\omega_1$ is in the
vicinity of $\omega_c$. The equation does not take account of the
effect of collisions on power absorption. A graph of this
equation is shown in Figure 4-2. The secondary maxima can be
seen only at low pressure when the magnetic and electrical
fields are perfectly homogeneous. Thus they are usually
missing.

Reactive collisions between ions and molecules render ICR
spectrometry chemically interesting, but such collisions, as
well as nonreactive ones, complicate the ICR theory. Under the
circumstances, it is fortuitous that equations such as the one
just given are often good approximations even though they do not
contain a collision term.

Three types of collision can be identified: (a) reactive
collision, in which there is transfer of both mass (one or more
atoms) and momentum; this type has been discussed throughout the
book, (b) nonreactive collision, in which only momentum is ·
transferred, and (c) charge-transfer collision, in which an
electron moves from the molecule to the cation. The transfer of
both mass and momentum are negligible. However since the ion is
neutralized and the molecule ionized, the effect is generally a
definite change in both the mass and the momentum of the ion.
For the special case of charge exchange between M and $M^+$, the

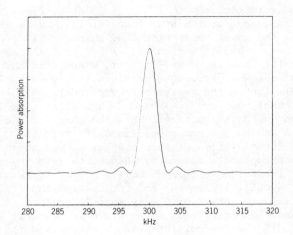

Figure 4-2.  Calculated ICR power absorption as a function of frequency.  (Figure courtesy of W. B. Nixon.)

mass of the ion does not change; this is the symmetrical charge-exchange reaction.

A power equation that takes account of collisions is[A35]

$$A(\omega_1) = (\varepsilon_{rf}^2 e^2 \tau/4m) \frac{\xi}{(\omega_1 - \omega_c)^2 + \xi^2}$$

$A(\omega_1)$ is the power drawn from the irradiating oscillator per ion, and $\xi$ is the collision frequency for momentum transfer.  We can easily show that this equation leads to a direct evaluation of $\xi$.  For simplicity let $R = \varepsilon_{rf}^2 e^2 \tau/4m$.  The power absorbed at $\omega_1 = \omega_c$ is given by

$$A(\omega_1 = \omega_c) = \frac{R\xi}{\xi^2} = \frac{R}{\xi}$$

The value of $\omega_1 - \omega_c$ that corresponds to the half-height is obtained from

$$\frac{R}{2\xi} = \frac{R\xi}{(\omega_1 - \omega_c)^2 + \xi^2}$$

and this leads immediately to $\xi = \omega_1 - \omega_c$ at half-height. The collision frequency for momentum transfer is thus readily derived from the power absorption line shape. This is true at high pressures,[2] when an ion undergoes many collisions while it is in the cell, that is, when $\xi^{-1} << \tau$, where $\tau$ is the drift time through the analyzer. At low pressures, when the likelihood of collision in the analyzer is very small, the linewidth depends inversely on $\tau$.[A9,C57,C25,A45] The drift time depends inversely on the drift voltage. Thus the sharpest ICR signals are obtained at low pressures and low drift voltages.

In the first section of this chapter it was shown that whatever interrupts the resonance phenomenon sets the signal linewidth. If an ion suffers no collisions of any kind as it traverses the cell, its energy absorption from the radio-frequency field is interrupted only when it drifts out of the analyzer, and the signal will be as sharp as possible for that drift time. If the ion collides with a molecule while in the analyzer, it will be removed from resonance, and the collision frequency sets the linewidth. This principle has been applied to the autoionization of anions, $X^- \rightarrow X + e^-$. The absorption of power by the anion ends when it ejects the electron. The process is assumed to be unimolecular,

$$\frac{d(X^-)}{dt} = -k(X^-)$$

where $k$ is the first-order rate constant for the disappearance of $X^-$. If $k^{-1}$ is considerably shorter than the drift time, but long enough to allow the anions to enter the analyzer before losing their electrons, the reaction rate will determine the signal linewidth.[C17] In this way the autoionization lifetimes of $SF_6^-$ and $C_4F_8^-$ were found to lie in the vicinity of 500 and 200 $\mu$sec respectively.[3]

An alternate measurement of collision frequencies has been made in an ion-pulsing experiment,[C27] which we now consider. The instantaneous power absorbed by a single ion at resonance in the absence of collisions was found earlier (Chapter 1, Equation (7)). The instantaneous power absorption at $\omega_1 = \omega_c$ for a number $N^+$ of ions being removed from resonance by collisions at a rate $\xi$ $s^{-1}$ is[C27]

$$A(t) = \frac{N^+ \varepsilon_{rf}^2 e^2}{4m\xi}(1 - e^{-\xi t}) \tag{5}$$

This equation is most clearly understood if we choose the time between collisions, $\xi^{-1}$, as the unit of time for ions drifting through the analyzer. Suppose the analysis time t is quite short compared to the time between collisions; $t << \xi^{-1}$ and the exponential term can be expanded in the series

$$e^{-\xi t} = 1 - \xi t + \cdots$$

where only the first two terms need to be retained. Equation (5) is then

$$A(t) = \frac{N^+ \varepsilon_{rf}^2 e^2 t}{4m} \qquad (6)$$

Equation (7) of Chapter 1 lacks the $N^+$, but is otherwise identical to Equation (6). This is as it should be, because collisions were ignored in the derivation of the former.

The second special case of interest is that of $t >> \xi^{-1}$, for which $e^{-\xi t}$ becomes small compared to unity. Equation (5) thus becomes[4]

$$A(\xi) = \frac{N^+ \varepsilon_{rf}^2 e^2}{4m\xi} \qquad (7)$$

This expression does not depend on the time factor. The ions are losing energy by collision at the same rate as they are absorbing it from the irradiating oscillator. This is the steady-state condition, and power absorption is said to be saturated.

The collision frequency for momentum transfer $\xi$ is found as follows. The value of A(t) for a known t is measured at a low pressure, where there are essentially no collisions, and Equation (6) is correct. This yields a value of $(N^+ \varepsilon_{rf}^2 e^2 / 4m)$. A second measurement of A is made at high pressure, so that Equation (7) is correct. This measured value of $A(\xi)$ and the set of constants evaluated in the previous experiment allow Equation (7) to be solved for $\xi$.

The observed instantaneous power absorption as a function of time is in good agreement with the results calculated for nonreactive collisions. This is shown in Figure 4-3.

Because the collision frequency depends on the molecular number density n (related to the pressure by n = (P/kT), where k is the Boltzmann constant), the pressure-independent rate constant for momentum transfer is usually reported, and is defined by $k = (\xi/n)$. For $N_2^+$ in $N_2$, $k = (0.67 \pm 0.01) \times 10^{-9}$ $cm^3$ molecule$^{-1}$ s$^{-1}$.

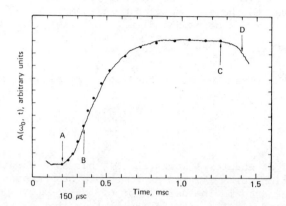

Figure 4-3. Experimental instantaneous power absorption curve versus theoretical (dots) for $N_2^+$ ions in
$N_2$. The ion packet starts to enter the analyzer region
at point A and has fully entered at point B. The ion
pulse width is 150 μsec. Calculated point of ion
packet departure is point D; the actual departure
begins at point C. The pressure is $2.2 \times 10^{-4}$ torr,
and the electric field strength of the detecting
oscillator is as low as possible, 0.0104 V/cm, so as
to impart minimal energy to the ions. (Reprinted, by
permission, from *J. Chem. Phys.* **55**, 2146 (1971), Figure 1.)


## THE LANGEVIN AND GIOUMOUSIS-STEVENSON EQUATIONS

An equation derived early in the present century by the French
physicist Paul Langevin describes the attraction between an ion
and a nonpolar molecule. Let e be the charge on the ion, α the
polarizability of the molecule, E the electric field intensity
around the ion, r the center-to-center distance between the ion
and the molecule, and μ the induced dipole moment of the
molecule. Langevin first writes:

$$\mu = \alpha E$$

He continues, "An ion of unit electronic charge sets up an
electric field of intensity $e/r^2$ at a distance r, so that the

force of attraction between it and the molecule is[5]:

$$F = \mu \frac{dE}{dr} = \alpha E \frac{dE}{dr} = \alpha \frac{e}{r^2} \frac{d}{dr}\left(\frac{e}{r^2}\right) = \frac{-2\alpha e^2}{r^5} \tag{8}$$

The attractive force thus varies as the inverse fifth power of the distance. The work done by this force as the distance decreases from infinity to r is:

$$w = \int_r^\infty F dr = \frac{-\alpha e^2}{2r^4} \tag{9}$$

(End of quotation)[6]  The potential energy (energy of attraction) of the ion-nonpolar molecule system when the two are a distance r apart is thus:

$$V = \frac{-\alpha e^2}{2r^4} \tag{10}$$

This is the Langevin equation.[7]  As written, it assumes electrostatic units. More convenient units are obtained by using K, the constant that appears in Coulomb's law whenever electrostatic units are not used. The equation becomes:

$$V = \frac{-K\alpha e^2}{2r^4} \tag{11}$$

where $K = 14.4 \text{ eV·Å}/e^2$, the polarizability is expressed in $\text{Å}^3$, e is in electronic charge units, and r in Å. As an example, for a singly charged ion 4 Å from a molecule whose polarizability is 3 $\text{Å}^3$, the potential energy is -0.08 eV. The average translational energy of a molecule at room temperature is 0.04 eV, so the attractive force is appreciable if the ion and molecule are close to each other.

When an ion and a molecule are in contact according to the collision diameters of kinetic theory, the contribution of the London dispersion energy, which varies as $r^{-6}$, becomes important. For $Cs^+$ colliding with He or Ne, this contribution to the potential energy is comparable to that of Equation (11), whereas even for $Li^+$ (which has a very small polarizability) in collision with the permanent gases, the potential energy is at least 15% more negative than the value calculated from Equation (11).[8]  However the perturbation of the electron clouds of the ion and molecule at short range is great enough that one cannot be naive in the use of such terms to obtain "correct" potential energies. This conclusion has emerged from certain scattering experiments.

The dependence of the potential energy on the polarizability provides an interesting *caveat* in the logic of ion-
molecule reactions. Although the systems $A^+ + B$ and $A + B^+$ have
identical nuclei, the same number of electrons, and very similar
geometries, the Langevin equation shows that they will not be
energetically equal at the moment of reaction unless A and B
have the same polarizability. A further criterion is that they
have the same ionization energies. Thus the two reactions will
in general have different rates, and sometimes even different
products.

The Langevin equation can be related to the measurable
parameters of ion-molecule reactions in terms of collision
dynamics.[9] The rate constant k is a function of the reaction
cross section $\sigma$[10] and the relative velocity g according to

$$k = \int_0^\infty f(g)\sigma(g)g \; dg$$

Here f(g) is the relative velocity-distribution function. The
equation is valid only when f(g) and the average internal energy
of the molecules are invariant during the measurement. These
are two rather severe criteria, given the way in which ion-
molecule reaction studies are usually conducted.[11] Details of
the energy problem are considered later.

Ideally, f(g) is the Maxwell distribution, but this may not
be assumed. If a reaction occurs at almost every ion-molecule
collision, there is a net depletion of fast-moving molecules and
ions, and the distribution changes. This problem and the one
concerning internal energy can be overcome by the addition of an
inert buffer gas whose sole purpose is to maintain thermal
equilibrium via nonreactive collisions. Its pressure must be
such that most of the collisions of reactants and products are
with the buffer gas.

We shall now try to "picture" the cross section $\sigma$. In
order to represent the relative velocity as simply as possible,
one of the collision partners will be at rest. When it becomes
convenient to make the choice, we will say that the molecule is
moving with respect to the ion.[12]

Consider first the case of the hard-sphere collision model
of kinetic theory. At best, it is realistic only for atoms and
molecules, not for ions. Figure 4-4(A) shows a molecule A at
the left that moves in a straight line past a second molecule B
without collision. As shown, the *impact parameter* b is the
distance from the center of B to the projected path of the
center of A when there is no force between them. Let $r_A$ and $r_B$

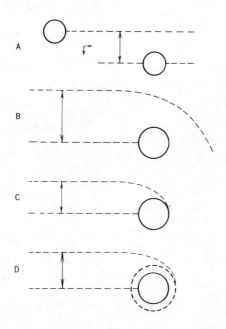

Figure 4-4.   (A) Molecule A moves past molecule B.   The
impact parameter b is such that they do not collide.
The absence of any deviation in the path of molecule A
fits the hard-sphere model for b > $r_A$ + $r_B$.   In Figures
4-4(B-D), Molecule A is regarded as a point mass
(polarizable, however!), and the size of both A and B
is assigned to the ion B.   This is the custom in draw-
ing Langevin trajectories and makes them easier to
follow.   These three diagrams illustrate the Langevin
model.   (B) The impact parameter is greater than its
critical value $b_C$, so there is no collision.   (C) The
impact parameter b is less than the critical value $b_C$,
and the molecule collides with the ion.   (D) The impact
parameter is at the critical value, and the molecule
moves into a fixed orbit at b = $b_C/\sqrt{2}$.

     Figures 4-4(B-D) assume that the relative velocity
is not large.   Equation (14) predicts that the critical
impact parameter becomes arbitrarily small for large

be the radii of the respective molecules.  The critical value of
b for collision is $b_C = r_A + r_B$.  The two spheres will collide
for all values of b such that $0 \leq b \leq b_C$.  When $b = 0$, the
collision is head-on.  Let A move in the x direction, as shown.
The collision cross section, $\sigma$, is the area of the circle in the
yz plane given by[13]:

$$\sigma = \pi (r_A + r_B)^2 = \pi b_C^2 \tag{12}$$

According to the hard-sphere model, there is no potential
energy between A and B except at the moment of contact, when it
has an infinite positive value.  Hard spheres do not compress
each other in collision.  If we take the Langevin model as a
more realistic description of $V(r)$ for an ion-molecule system,
our conclusions about what happens as the two come together will
be entirely different, but explicable in terms of the parameters
b and $\sigma$.

Two kinds of trajectory are possible as a molecule moves in
the vicinity of an ion.  An "open" trajectory is shown in
Figure 4-4(B).  The path of the molecule is deflected toward the
ion because of the negative potential energy, but the molecule
leaves the vicinity without a collision.

A "closed" trajectory is shown in Figure 4-4(C).  The
molecule is drawn into contact with the ion.  This situation
leads to ion-molecule reactions.

In theory, there exists a special case, a closed trajectory
that does not lead to ion-molecule contact.  It arises when the
attraction between ion and molecule is great enough to draw the
molecule into a fixed orbit near the ion but not to bring them
together.  This leads to the definition of a critical impact
parameter $b_C$ for systems that obey Equation (10).  Langevin
showed that it is

$$b_C = \left(\frac{4e^2\alpha}{\mu g^2}\right)^{1/4} \tag{13}$$

where $\mu$ is the reduced mass of the pair and $g$ the relative
velocity.  A molecule for which $b = b_C$ will approach the ion
until $b = b_C/\sqrt{2}$, the radius of the fixed orbit shown in Figure
4-4(D).  Total energy of the system is of course conserved.
Because the potential energy becomes more negative as r
decreases, the speed of the molecule must increase as it
approaches the ion.  At $b_C/\sqrt{2}$ the attractive force is able to

_____

values of the relative velocity $g$, a possibility that
is ignored in these diagrams.

maintain the centripetal acceleration of the molecule, but no longer to increase it, so the molecule goes into a fixed orbit.

The cross section for a collision between a nonpolar molecule of polarizability $\alpha$ and an ion of charge e is, according to the Langevin model,

$$\sigma(g) = \pi b_C^2 = (2\pi e/g)(\alpha/\mu)^{1/2} \tag{14}$$

We thus obtain a simple expression for the collision cross section.[14] It is influenced by the molecule only through $\alpha$ and $\mu$.

Because the activation energy of ion-molecule reactions is frequently near zero, it is often assumed that every collision leads to reaction.[15] This further requires the absence of steric or orientation factors. To the extent that these criteria are met, Equation (14) may be regarded as the reaction cross section.

An alternate form of Equation (14) is obtained by noting that $E_{tr} = \mu g^2/2$, where $E_{tr}$ is the relative translational energy of the two-particle system. Replacement of $\mu g^2$ within the parentheses gives:

$$\sigma(E_{tr}) = \pi e(2\alpha/E_{tr})^{1/2} \tag{15}$$

Reactions fit the model when their cross sections show this $E_{tr}^{-1/2}$ dependence.[16]

The cross section depends inversely on the relative velocity of the molecule and the ion. The force of attraction between the pair must accelerate the molecule toward the ion[17] if there is to be a reaction, and the magnitude of this acceleration depends on the time the ion spends near the molecule, thus on $g^{-1}$. When g is large, the ion must pass close to the molecule if it is to move into a closed trajectory, while for slow molecules, $b_C$ and hence $\sigma(g)$ are greater.

If each collision is reactive, the collision frequency will be the rate of reaction. The rate of a bimolecular reaction is $R = -kN_1N_2$, where k is the rate constant, and $N_1$ and $N_2$ are the concentrations, or more conveniently the number densities of the two kinds of molecule. The collision rate between particles of types 1 and 2 depends on the product of these number densities, a cross section $\sigma$, and the relative velocity g. Having argued that the rates of collision and reaction are equal leads to a reaction-rate constant[18] given by:

$$k = g\sigma(g) = 2\pi e(\alpha/\mu)^{1/2} \tag{16}$$

This is the equation of Gioumousis and Stevenson,[19] who obtained it assuming a Maxwellian velocity distribution for both ions and

molecules.  Our derivation has been greatly simplified by
incorporating the average relative velocity throughout, rather
than starting with distribution functions.

If the rate of disappearance of a reactant ion obeys Equation (16), it will be independent of the translational energy of
the reactant ions.  Such a reaction would produce a double-
resonance signal if one of two conditions prevailed:  (a)
reactant ions were being removed from the cell and (b) the
reactant ion were forming two sets of products via reactions
with rate constants $k_1(E)$ and $k_2(E)$ whose sum was not a function
of the energy.  An example appears in Figure 4-8.

The Langevin-Gioumousis-Stevenson model thus predicts both
a numerical value of k for appropriate reactions and the absence
of an ion translational energy effect on k.  This is not always
fully recognized.[20]

Equation (16) has brought us at last to a directly measurable quantity.  Observed rate constants fit this equation
reasonably well for some simple ion-molecule reactions, as
originally reported by Gioumousis and Stevenson.  Others have
made similar observations, although the limitations of the
Langevin equation, as revealed by departures from the
Gioumousis-Stevenson equation, have been the subject of a
number of experiments.[21]

The Langevin model does not deal with the case of molecules
that have permanent dipole moments.  We consider such a model
later in this chapter.

REACTION RATES

The stoichiometry and the rate of a reaction are its two most
important characteristics.  The former presents only occasional
problems to the ICR investigator.

The measurement of rate constants has been of great
interest since the inception of ICR chemistry.  Early methods on
the unmodified instrument were difficult and not very satisfactory.  They are, however, instructive.

*Rate Constants by an Early Method in the Drift Cell*[E9]

The calculation of rate constants was first done from conventional drift-cell data as follows.  In the reaction
$P^+ + N \xrightarrow{k} S^+$ + neutral product, $P^+$ is the primary ion, formed
by the electron beam, and $S^+$ is a secondary (product) ion.  The
reaction occurs in the cell of a conventional instrument.  The
respective ion currents of primary and secondary ions will be

denoted by P(t) and S(t); $P_0$ is the current of primary ions formed at the electron beam. $P^+$ undergoes no other reactions.

If the course of the primary ions formed during a differential time could be followed, it would be seen that

$$\frac{-dP(t)}{dt} = knP(t)$$

The reaction is bimolecular, and n is the number density of neutral reactants. In a conventional bimolecular reaction n would decrease with time. Here it does not, for two reasons. First, the reactant gas is continually entering the cell; second, so few molecules are ionized that this does not appreciably reduce their numbers. Thus the mathematics is that of a first-order reaction, for which

$$P(t) = P_0 e^{-nkt} \tag{17}$$

Since the reaction forms as many ions as it consumes, and since they drift at the same speed for a given value of the magnetic flux density,

$$P(t) + S(t) = P_0$$

This requires that

$$S(t) = P_0(1 - e^{-nkt}) \tag{18}$$

The same result for S(t) comes from integration of $dS/dt = knP(t)$, treating n as a constant, and inserting the expression for P(t) given above.

When a primary ion undergoes many collisions during its passage through the cell, the power it absorbs from the marginal oscillator as the oscillator angular frequency $\omega_1$ sweeps across the resonant angular frequency of the primary ion $\omega_p$ is given by

$$A(\omega_1) = \frac{n_p^+ \varepsilon_{rf}^2 e^2}{4m_p} \frac{\xi_p}{(\omega_1 - \omega_p)^2 + \xi_p^2}$$

Here $n_p^+$ is the number of primary ions in the cell, a quantity that decreases with distance from the electron beam but that does not change with time; $m_p$ is the mass of the primary ion, and $\xi_p$ is the collision frequency for momentum transfer for the primary ion. At $\omega_1 = \omega_p$ the power absorption becomes:

$$A(\omega_1 = \omega_p) = n_p^+ \varepsilon_{rf}^2 e^2 / 4m_p \xi_p \tag{19}$$

The variation of $n_p^+$ with distance from the electron beam is the same as the variation of $P(t)$ with time, still supposing that we could follow the course of primary ions formed during a differential time.

Using Equation (19), the power absorption *per primary ion* at resonance can be written $A_p = C_p/m_p$, where $C_p = \varepsilon_{rf}^2 e^2/4\xi_p$. The signal intensity for the primary ions $P^+$ is, according to Equation (13) of Chapter 1,

$$I_p = \int_{\tau'}^{\tau} \frac{C_p}{m_p} P_0 e^{-nkt}\, dt = \frac{C_p}{m_p} P_0 \int_{\tau'}^{\tau} e^{-nkt}\, dt$$

where $\tau'$ is the time the ion enters the analyzer region, and $\tau$ the time it leaves it.[22]  Integration gives

$$I_p = \left(C_p P_0/m_p nk\right)\left(\exp(-nk\tau_p') - \exp(-nk\tau_p)\right) \qquad (20)$$

The subscript is added to the times $\tau'$ and $\tau$ because they depend on the magnetic field at which $\omega_p = \omega_1$. Equation (20) states that the primary ion signal is proportional to the difference between the number of such ions when they enter the analyzer and the number remaining when they leave it.

An analogous treatment leads to

$$I_s = \frac{C_s P_0}{m_s nk}\left(nk(\tau_s - \tau_s') - \exp(-nk\tau_s') + \exp(-nk\tau_s)\right) \qquad (21)$$

for the current of secondary ions.

Our task is to extract from Equations (20) and (21) an expression for the rate constant in terms of measurable quantities. The drift times of $S^+$ can be written in terms of the drift times of $P^+$ by use of the basic cyclotron equation $\omega_c = e(B/m)$, the drift velocity equation $v_{drift} = (\varepsilon_{drift}/B)$, and the inverse dependence of the drift time on $v_{drift}$. It follows that the drift time of an ion in resonance when B is being swept is proportional to its mass. In terms of the ions of interest,

$$\frac{m_p}{\tau_p} = \frac{m_s}{\tau_s}$$

The same is obviously true for $\tau'$.  Thus

$$\tau_s = \frac{m_s}{m_p} \tau_p \quad \text{and} \quad \tau_s' = \frac{m_s}{m_p} \tau_p' \tag{22}$$

It is usually the case that nkt << 1. All of the exponential terms can be expanded using the series

$$e^{-x} = 1 - x + x^2/2! - \cdots$$

For $I_p$ the expansion goes as follows:

$$I_p = \frac{C_p P_0}{m_p nk}\left((1 - nk\tau_p' + n^2 k^2 \tau_p'^2/2) - (1 - nk\tau_p + n^2 k^2 \tau_p^2/2)\right)$$

$$I_p = \frac{C_p P_0}{m_p nk}\left(nk(\tau_p - \tau_p') + \frac{n^2 k^2}{2}(\tau_p'^2 - \tau_p^2)\right)$$

$$= \frac{C_p P_0}{m_p}\left((\tau_p - \tau_p') + \frac{nk}{2}(\tau_p'^2 - \tau_p^2)\right)$$

Use of the identity

$$\tau_p'^2 - \tau_p^2 \equiv (\tau_p' - \tau_p)(\tau_p' + \tau_p) \equiv - (\tau_p - \tau_p')(\tau_p + \tau_p')$$

leads to

$$I_p = \frac{C_p P_0}{m_p}(\tau_p - \tau_p')\left(1 - \frac{nk}{2}(\tau_p + \tau_p')\right) \tag{23}$$

An analogous expansion of Equation (21) leads to

$$I_s = \frac{C_s P_0}{m_s}(\tau_s - \tau_s')\frac{nk}{2}(\tau_s + \tau_s') \tag{24}$$

The number of variables in Equations (23) and (24) can be reduced by inserting the relations in Equation (22) into Equation (24).

$$I_s = \frac{C_s P_0}{m_s}\left(\frac{m_s}{m_p}\tau_p - \frac{m_s}{m_p}\tau_p'\right)\frac{nk}{2}\left(\frac{m_s}{m_p}\tau_p + \frac{m_s}{m_p}\tau_p'\right)$$

$$= \frac{C_s P_0}{m_p}(\tau_p - \tau_p')\frac{nk}{2}\frac{m_s}{m_p}(\tau_p + \tau_p')$$

$$I_s = \frac{C_s nkm_s}{2m_p{}^2} P_0 (\tau_p - \tau_p')(\tau_p + \tau_p') \qquad (25)$$

It is seen in the definition of $C_p$ that it depends inversely on $\xi_p$. $C_s$ likewise depends on $\xi_s{}^{-1}$. Equations (23) and (25) can be solved for $C_p$ and $C_s$ respectively, and these can be replaced by $\xi_p$ and $\xi_s$.

$$C_p = \frac{I_p m_p}{P_0(\tau_p - \tau_p')\left(1 - \frac{nk}{2}(\tau_p + \tau_p')\right)}$$

$$C_s = \frac{2m_p{}^2 I_s}{P_0 nkm_s(\tau_p - \tau_p')(\tau_p + \tau_p')}$$

$$\frac{C_s}{C_p} = \frac{\xi_p}{\xi_s} = \frac{2m_p{}^2 I_s P_0 (\tau_p - \tau_p')\left(1 - (nk/2)(\tau_p - \tau_p')\right)}{P_0 nkm_s(\tau_p - \tau_p')(\tau_p + \tau_p') I_p m_p}$$

Canceling the common factors in numerator and denominator and then multiplying by both denominators gives

$$nkm_s \xi_p(\tau_p + \tau_p')I_p = 2m_p \xi_s I_s\left(1 - (nk/2)(\tau_p + \tau_p')\right)$$

This can readily be solved for the rate constant, which is

$$k = \frac{2m_p \xi_s I_s}{n(m_p \xi_s I_s + m_s \xi_p I_p)(\tau_p + \tau_p')} \qquad (26)$$

Either n or the drift times $\tau_p$ and $\tau_p'$ can be varied if it is desired to measure k graphically.

One problem in the use of Equation (26) is the apparent need for collision frequencies for momentum transfer. They can be calculated[C46] or obtained from other measurements.[23] Rate constants calculated without them do not appear to be seriously in error, however. Equation (26) has been used[E9],[C6] and quoted[C32] without collision frequencies.[24]

A second problem is that the values of $\tau$ and $\tau'$ must be calculated from the drift voltages and the known dimensions of the cell. It is better to measure ion-residence times directly, as can be done in modified cells.

A restraint on the utility of Equation (26), at least in theory, is that the power Equation (19) is the limiting case for many collisions, and systems in which the reactant undergoes

only one reaction at high pressures are the exception rather
than the rule.  The calculated rate constants do not seem to
depend greatly on such approximations, however.

The derivation of Equation (26) can be generalized to the
case of more than one simultaneous reaction.

As the subject has been developed thus far, the rate constants obtained from the conventional instrument seem rather
inaccessible, both in terms of the theory and the calculational
method.  Further evidence of this appears in Appendix 1, which
gives some details of an algorithm for the calculation of rate
constants for a general reaction system that extends to tertiary
ions.

## An Approximation for Drift Cell Calculations[C46]

An approximate means of calculating rate constants from the data
provided by conventional cells begins with a series expansion of
the ion-current equation.  Equation (17) becomes[25]

$$P(t') = P_0 (1 - nkt')  \tag{27}$$

The ensuing mathematics is considerably simplified.  The equation for the resonant power absorption per ion that is used in
this approximation is

$$A(t) = \frac{\varepsilon_{rf}^2 e^2 t}{4m}$$

By Equation (13) of Chapter 1, the signal intensity of primary
ions is

$$I_p = \int_0^{\tau_p} \frac{\varepsilon_{rf}^2 e^2 t P_0}{4m_p} \left(1 - nk(\tau_p' - t)\right) \, dt$$

$$= \frac{\varepsilon_{rf}^2 e^2 P_0}{4m_p} \int_0^{\tau_p} (t - nk\tau_p' t - nkt^2) \, dt$$

$$I_p = \frac{\varepsilon_{rf}^2 e^2 P_0}{4m_p} \frac{\tau_p^2}{2} \left(1 - nk(\tau_p' + \frac{2}{3}\tau_p)\right)  \tag{28}$$

The power absorbed by a secondary ion is more difficult to
calculate.  For the $i^{th}$ secondary ion it is

$$I_{si} = \frac{\varepsilon_{rf}^2 e^2 \cdot}{4m_{si}} \frac{P_0}{2} nk_i \tau_{si}^2 \left(\tau_{si}' + \frac{\tau_{si}}{3}\right) \tag{29}$$

There will be as many equations of the form (29) as there are kinds of secondary ions formed by the reactant primary. Taking all such reactions into account, Equations (28) and (29) can be solved for

$$k_i = \frac{I_{si}}{I_p C_{psi} + B(I_{si} + \sum_{j \neq i} I_{sj} C_{sjsi})} \cdot \frac{1}{n} \cdot \frac{1}{(\tau_p' + \tau_p/3)} \tag{30}$$

in which the factor $B = (\tau_p' + 2\tau_p/3)/(\tau_p' + \tau_p/3)$ depends on the drift voltages and the lengths of the source and analyzer. If the signals are produced in the usual fashion by sweeping the magnetic field,

$$C_{psi} = \left(\frac{m_{si}}{m_p}\right)^2 \quad \text{and} \quad C_{sjsi} = \left(\frac{m_{si}}{m_{sj}}\right)^2$$

Rather than calculate a value of $k_i$ directly from Equation (30), it is better to determine it from data at several pressures. Multiplication of Equation (30) by n, the number density of the gas, gives an equation in the standard linear form, with $k_i$ the slope and n the independent variable.

Similar arguments and equations have been developed for the measurement of rate constants from ion-current data.[C22]

Direct measurements of the residence times[26] of ions in conventional and modified drift cells have increased the confidence in the accuracy of the rate constants obtained in ICR measurements.[A20,C31] The measurement of rate constants has become routine with the development of new cells.

*Rate Constants Measured in a Four-Section Cell*[E39,C20,C31]

A four-section cell gives good control over ion paths and permits a simple calculation of rate constants. For the reaction $P^+ + N \xrightarrow{k} S^+$ + neutral product, we let $I_{p+}$ be the signal intensity of the primary ion, and TIC the total ion current. The rate constant can then be found graphically from the equation

$$\ln \frac{I_{p+}}{TIC} = -kn\tau \tag{31}$$

if n, the number density of the neutral N, is varied.  The slope
is $k\tau$.  The residence time is calculated by the use of

$$\tau = \tau_s + \tau_{rx} + \frac{\tau_a}{2}$$

where s, rx, and a refer respectively to the source, reaction,
and analyzer sections of the cell, which is shown in Figure 4-5.
Division by 2 in the last term takes account of the detection of
ions throughout the analyzer, and selects the time when they
reach its midpoint as being proportional to the signal intensity.  Mass-dependence corrections do not appear in Equation
(31) because they would affect the intercept, but not the slope
of a graph of the equation.

### Rate Constants from Trapped Ion Cells[A19]

Equation (17) suggests that the measurement of rate constants
would be straightforward if P(t) data were available.  This is
the case when kinetic studies are done in ion trapping or
storage cells, as discussed in Chapter 1.  In each case the
signal intensity (occasionally the ion current) of an ion is
measured after the reaction has been in progress for a known
time, typically no longer than 0.1 s.  The detection time is
also known, so signal intensity is directly proportional to ion
current.  Rate constants for the disappearance of primary ions
and the appearance of secondary ions are readily calculated.

Because the number of ions can vary slightly during an
experiment, the total signal intensity $\Sigma I$ is sometimes used to
normalize the data.  Equation (17) becomes:

$$I(t) = I_0 e^{-nkt}$$

or

$$\ln \frac{I(t)}{I_0} = -nkt$$

$I_0$, the intensity at zero time of the ion under study, is
replaced by $\Sigma I$, of which it is a constant fraction.  This has no
effect on the calculated value of k.

As an example we calculate the rate constant for the disappearance of $H_2O^+$ by reaction with $H_2O$, as shown in Figure
4-6.[C61]  The number density at $1.06 \times 10^{-6}$ torr is calculated
from the ideal-gas equation to be $3.41 \times 10^{10}$ molecules/cm$^3$ at a
temperature of 300 K.[27]  The slope as read from the extremes of
the $H_2O^+$ line is

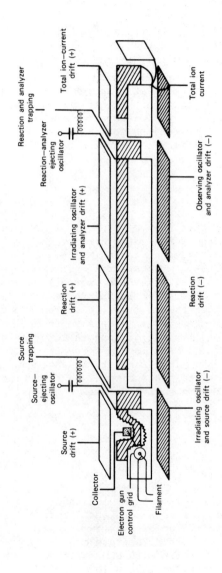

Figure 4-5. A four-section drift cell. As shown, it can be used for ion-ejection studies. (Reprinted, by permission, from *J. Amer. Chem. Soc. 94*, 3748 (1972), Figure 1.)

146

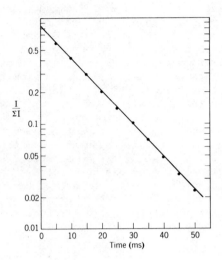

Figure 4-6.   Kinetic study of the reactions of $H_2O^+$
with $H_2O$ at a pressure of $1.06 \times 10^{-6}$ torr.

$$\text{slope} = \frac{\ln 0.82 - \ln 0.020}{0 - 0.0525 \text{ s}} = -70.73 \text{ s}^{-1}$$

The rate constant is

$$k = -\frac{\text{slope}}{n} = 2.07 \times 10^{-9} \text{ cm}^3 \text{ molecule}^{-1} \text{ s}^{-1}$$

Many rate constants, for the formation of secondary ions as well
as the disappearance of primary ions, have been measured in this
way.  Such an operation, known as the "reaction-kinetic mode"
(RKM), is available on the new commercial instruments, and ICR
is sure to be a prolific source of the rate constants of gaseous
ion-molecule reactions.

Values of absolute rate constants measured by ICR spectrometry have been tabulated.[A35]

ENERGY AND THE RATES OF ION-MOLECULE REACTIONS

Most of the present chapter deals in some way with the role of
energy in ion-molecule reactions, and comments on the subject
are scattered throughout the earlier chapters.  One section of
the Bibliography suggests the variety of dynamic and thermochemical energies that influence such reactions.  The present
section will be limited to a descriptive review of the energy of
reactant ions and its influence on reactions, omitting experiments such as the collisional stabilization of excited product
ions.  The kinetic energy available for reaction will be
discussed before turning to reaction rates.

In certain ICR experiments a radiofrequency field is used
to accelerate the reactant ions to an approximately known energy
prior to reaction.  In this way the dependence of rate on
kinetic (translational) energy can be studied.  A reaction is an
example of a very inelastic collision except for the rare case
of $\Delta E = 0$.  Ignoring entropy, kinetic energy is converted into
potential energy if product bonds are stronger than reactant
bonds.  Kinetic energy is not conserved, although momentum must
be conserved.  The requirement that there be some momentum, and
hence some kinetic energy after reaction, leaves only a fraction
of the reactant-ion kinetic energy for conversion to potential
energy.  This fraction is the available kinetic energy, and is
given by

$$E_a = E_i \left(\frac{M}{m + M}\right)$$

where $E_i$ is the kinetic energy of the ion before collision, and
M and m are the masses of the molecule and the ion respectively.
The equation is derived using a center-of-mass coordinate
system,[28] and $E_a$ is usually denoted $E_{CM}$ when describing ion-
molecule reactions.  When this correction is not made, $E_i$ is
often referred to as the "laboratory (lab) energy."


*Rate Constants As a Function of Reactant Ion Energy*

The effects of ion energy on the reaction-rate constant fall
into two categories:  (a) ion kinetic (translational) energy and
(b) ion internal energy.  The first four examples illustrate the
former.  Figure 4-7[B45] shows data for four reactions between
$C_3H_6^+$ and $C_3D_6$.  The products are $C_5(H,D)_9^+$ and $C(H,D)_3$ in every
case, differing only in the distributions of H and D.  The
energy dependence is the same for the four isotopically different reactions.  A decrease in reaction rates with increasing ion

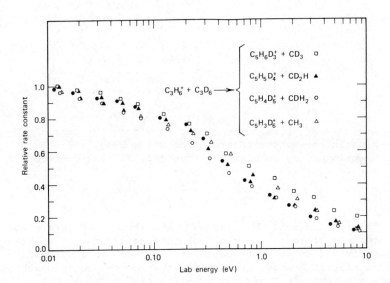

Figure 4-7.  Dependence of the relative rates of propene reactions on ion kinetic energy. (Reprinted, by
permission, from *Adv. Electron. Electron Phys. 34*, 223
(1973), Figure 22.)

kinetic energy is a very common observation.

Data for the reaction $Ar^+ + D_2 \rightarrow ArD^+ + D$ appear in Figure
4-8.[C20]  The signal intensity of the $ArD^+$ ion relative to the
sum of the two ion signal intensities is shown.  Its constancy
shows that the rate constant does not change over a 100-fold
variation in the kinetic energy of the $Ar^+$ ion.  This is a good
example of a reaction obeying the Langevin equation.  The rate
constant was measured as $9 \times 10^{-10}$ $cm^3$ $mol^{-1}$ $s^{-1}$.

The rate of the reaction $H_2^+ + He \rightarrow HeH^+ + H$ has been
studied by several methods, with generally good agreement among
the results.  Rate constants obtained in a four-section ICR cell
are shown in Figure 4-9.  The $H_2^+$ was formed by electron impact
at 50 eV, and undoubtedly received some vibrational energy in
the process.  Rate constants that go through a maximum are not

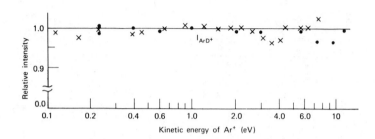

Figure 4-8. Dependence of the rate constant on ion kinetic energy for the reaction $Ar^+ + D_2 \rightarrow ArD^+ + D$. (Reprinted, by permission, from *Int. J. Mass Spectrom. Ion Phys.* **4**, 165 (1970), Figure 9.)

unusual.

The reactions of $NH_3^+$ with $NH_3$ are shown in Figure 4-10.[C23] The ordinate scale is the rate constant as a function of kinetic energy divided by the rate constant for reaction 1 with only thermal kinetic energy. Rate constants for the charge exchange reaction between $NH_3^+$ and $NH_3$ appear along the lower curve. The middle curve shows the data for the proton-transfer reaction. Use of $^{15}N$ resolves the composite signal that would otherwise

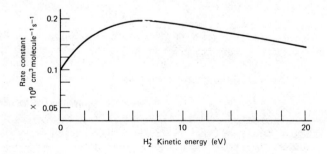

Figure 4-9. Dependence of the rate constant on ion kinetic energy for the reaction $H_2^+ + He \rightarrow HeH^+ + H$.

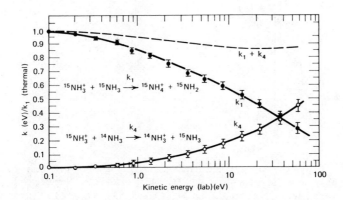

Figure 4-10. Kinetic energy dependences of the relative rates and total rate for the reactions of $NH_3^+$ with $NH_3$ measured in a mixture of $^{14}NH_3$ and $^{15}NH_3$. The ionization energy is 50 eV. (Reprinted, by permission, from *J. Chem. Phys.* **54**, 843 (1971), Figure 8.)

result for reactant and product $NH_3^+$. The dashed curve is the sum of the curves under it, and indicates that the total disappearance of $NH_3^+$ by reaction with the neutral molecule fits the Langevin model. Ionization was by 50 eV electrons.

In the four studies just discussed it should not be assumed that excess internal energy, if present in the ion, made no contribution to the reaction rates, but rather that its contribution was constant, whether negligible or not. This question will be taken up later.

The next three figures illustrate the effects of changes in the internal energy.[E22],[E44] Figure 4-10 showed the effect of kinetic energy on the ammonia system. Consequences of a variation in the electron impact energy on the reaction

$$NH_3^+ + NH_3 \longrightarrow NH_4^+ + NH_2$$

(Reaction 1 in Figure 4-10) are seen in Figure 4-11.[C61] When the electron energy exceeds the reaction threshold[29] by as little as 3 eV, the rate decreases by a factor of one-third. The reaction is more sensitive to an increase of 3 eV in electron energy in this energy range than to a comparable

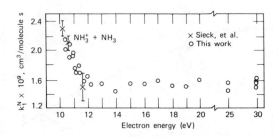

Figure 4-11.  Dependence on electron energy of the rate
of the reaction $NH_3^+ + NH_3 \rightarrow NH_4^+ + NH_2$.  (Reprinted,
by permission, from *J. Chem. Phys. 59*, 4742 (1973),
Figure 8.)

increase in kinetic energy.

Figure 4-12 reports a more detailed study of the same
general behavior.  The alkene M reacts with $M^+$ in a condensation
reaction that eliminates a methyl radical.[C24]  The electron
energy is varied from near the ionization potential to about
11 eV above it, and initially the rates fall sharply with
increasing energy, then less sharply until they become insensitive to the electron energy.

If these rate constants were decreasing due to a change in
electronic energy states of the ions, we would expect no difference in the $C_2H_4$ rates until $E_e$ - IP $\geq$ 1.87 eV, the first
electronic excited state of $C_2H_4^+$.  Figure 4-12 shows that this
is clearly not the case.  It is concluded that the decrease in
rates is due to vibrational excitation of the ion as it is
formed at the electron beam.  Nothing more precise can be said
when ionization is by electron impact because the spread of
electron energies in the ionizing beam is the same order of
magnitude as the larger vibrational quanta.

A modified instrument in which a small Dempster mass
spectrometer replaces the usual source region provides a third
example of the influence of internal energy on reaction rates,
and again the vibrational energy emerges as the channel through
which rates are influenced.[B46]  The reactions in this study are

$$D_3^+ + CH_4 \longrightarrow CH_4D^+ + D_2$$

and the isotopically different methyl cations that are formed from these reactants,

$$D_3^+ + CH_4 \longrightarrow CH_2D^+ + H_2 + D_2$$

$$D_3^+ + CH_4 \longrightarrow CH_3^+ + HD + D_2$$

The first reaction is exothermic by 1.0 eV and the last two are endothermic by 0.6 eV. A shift toward the last two reactions at the expense of the first one occurs whenever $CH_4D^+$ contains at least 1.6 eV of excess energy.

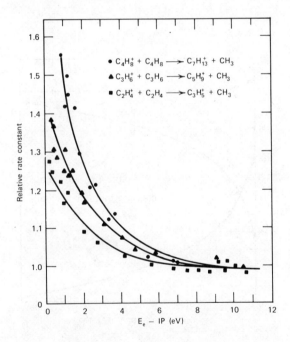

Figure 4-12. Relative rate constants as functions of the difference between the electron energy V and the molecular ionization potentials (energies) E. (Reprinted, by permission, from *J. Chem. Phys. 54*, 3651 (1971), Figure 1.)

The $D_3^+$ is formed in the ion source of a medium-pressure mass spectrometer by the familiar reaction

$$D_2^+ + D_2 \longrightarrow D_3^+ + D$$

The $D_3^+$ undergoes further collisions before leaving the source, and these reduce the excess energy it acquired at the time of its formation. Variation of the pressure alters the average number of collisions of each $D_3^+$ ion; the higher the source pressure, the lower the excess vibrational energy of the $D_3^+$.

The ions are drawn out of the source, and the $D_3^+$ is isolated in the usual mass spectrometric fashion, then decelerated to nearly thermal energy, and subsequently enters the ICR cell and reacts with methane. Figure 4-13 shows the correlation

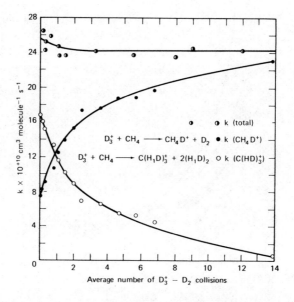

Figure 4-13. Variation of the total rate constant, $k(CH_4D^+)$ and $k\left(C(H,D)_3\right)^+$ as a function of the average number of $D_3^+$-$D_2$ collisions in the ion source. (Reprinted, by permission, from *J. Phys. B. Atomic and Molecular Physics* 8, 803 (1975), Figure 3.)

between reaction rates and the average number of $D_3^+$-$D_2$ collisions. The half-filled points correspond to the disappearance of $D_3^+$ and the filled and open points to the formation of $CH_4D^+$ and the methyl cations respectively. Although the near constancy of the total reaction rate for the disappearance of $D_3^+$ looks familiar, it has nothing to do with the Langevin model, which takes no account of the internal energy of the ions.

These observations of the effect of energy on ion-molecule reactions can be summarized by an analogy to solution reactions. As temperature affects the course of a typical synthetic reaction, higher temperatures favoring products different from those of lower temperatures, so also does the energy content of an ion affect its reactivity.

*Does ICR Spectrometry Produce Thermal Rate Constants?*[30]

Perusal of the literature appropriate to this question discloses a few reasons for doubt, plus a number of inconsistencies. For example, the phrase "near-thermal ion energies" is endemic. Readers are left to guess how near.[31] Another is the imprecise knowledge of the electron energy. The two forms of this problem are the natural energy distribution of electrons emitted by the hot filament, and failure to calibrate the electron energy against a known standard. The length of time a particular filament has been in use may even be a factor.

There is no reason to doubt that the molecules have a Maxwell-Boltzmann energy distribution at the cell temperature. For the reaction to be thermal, as the term is used here, the same is required of the ions.

For some reactions a small departure from ion isothermicity would be inconsequential. For kinetic energies, this is notably the case when the Langevin-Gioumousis-Stevenson theory is obeyed. It appears from the preceding discussion that internal energy often has a more immediate effect on reaction rates. Arguments pro and con will briefly be reviewed, starting with some that support the claim of thermal rate constants.

1. Comparisons of rate constants obtained by different techniques are frequently made in the literature, and the agreement between ICR values and others is usually satisfactory and sometimes excellent.[C61,C19,B45] The question of thermal rate measurements has also been raised with respect to some other methods for the study of ion-molecule reactions; the problem appears any time the ions are formed by electron impact.

2.  Excess energy is carried away ("relaxed") by non-
    reactive collisions.  This process can be enhanced by
    the addition of a nonreactive gas such as argon,
    helium, or nitrogen.  Used in this manner, they are
    called "buffer gases."  This has been done in a number
    of experiments; the presence of the gas has not altered
    the nature of the reactions (see the work by Dunbar[E56]
    and references cited therein).
3.  The strongest evidence that equilibrium can be attained
    in an ICR cell under certain conditions comes from
    determinations of equilibrium constants in a trapped
    ion cell.  Such a study of the reaction

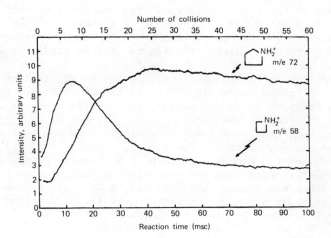

is shown in Figure 4-14.[B15]

Figure 4-14.   $(CH_2)_3NH_2^+$ and $(CH_2)_4NH_2^+$ signals as a
function of time in a 6.4:1 mixture of $(CH_2)_3NH$ and
$(CH_2)_4NH$ at 10 eV and approximately $1 \times 10^{-5}$ torr.  The
protonated species (m/e 58 and 72) were formed by proton transfer from the parent ions to the two neutral
molecules.  (Reprinted, by permission, from *J. Amer.
Chem. Soc. 93*, 4314 (1971), Figure 2.)

The equilibrium constant is 22 ± 2.  From the known neutral pressures and the reaction time it was calculated that the equilibrium was established after about 40 collisions.  It is difficult to survey the traces for the ions in the 60-100 ms region and argue against the establishment of both thermal and chemical equilibrium in this experiment.

Notwithstanding all of the above, there are several valid experimental factors that point the other way:

1. The cross section for ionization by electron impact is zero at the ionization potential.  Let us suppose that electrons having exactly 10 eV of kinetic energy strike a molecule whose ionization potential is exactly 10 V.  No ions will be formed.  In practice, one or another method of data treatment is used to calculate ionization potentials from electron-impact measurements made at voltages above the ionization potential.[32]  It is thus the nature of electron-impact ionization that the electron energy exceed the theoretical amount needed for ionization.  Photoionization has the advantage of a nonzero cross section at the ionization potential.

2. Neither electron impact nor photoionization is adiabatic except in special cases.  An adiabatic process, in this context, is one in which there is no vibrational excitation.  Adiabatic ionization potentials are obtained from spectroscopic studies.  An ion produced in an ICR cell will usually contain excess vibrational energy at the moment of formation.  This is true for both primary and secondary ions; the latter are almost always formed in exothermic reactions and will share the excess energy with the neutral.

3. A simple calculation will show why, in the extreme case, an excess of energy as small as 0.2 eV cannot be ignored.  For the reaction AB + CD → products, let A, B, C, and D be atoms.  The system has 12 degrees of freedom.  Each degree of freedom will gain an energy of $kT/2$, where k is the Boltzmann constant and T the absolute temperature.  The energy gained by the total system is 6 kT.  Setting this equal to 0.2 eV and solving for the temperature increase due to the excess energy, gives $390^0$.  Assuming randomization of the excess energy, the reactants are this much hotter than their environment.  A four-atom system is the extreme case because of the small number of degrees of freedom.  The effect does not seem negligible even for four times as many atoms.

No general answer to the thermal rate constant question can
be given.  It is less credible that a reaction will be thermal
if it occurs between a primary ion and the first molecule it
meets, than that thermal equilibrium is not attained in the
equilibrium-constant experiment discussed earlier.  Many ICR
studies examine reactions that fall between these extremes.  The
question is best answered by examining the experimental details
of each rate-constant measurement.

*Rate Theories*

This important topic is touched on here in order to point out
the recent contributions of ICR spectrometrists, who have
revived and extended one theory and spawned another.
    Consider the reaction

$$A^+ + BC \longrightarrow BA^+ + C \tag{32}$$

in which A, B, and C are the common central or "heavy" atoms (by
comparison with hydrogen) of ion-molecule chemistry.  Predictions about which reactions will occur are based on the electronegativities of these atoms.[C67]  The reactants and products can
also contain one or more hydrogen atoms; their presence does not
influence the prediction of reaction rates by this method.
    To illustrate this point, consider the following example.
Work must be done in order to move an electrical charge.  This
model states that a reaction involving the central atoms is
allowed if it requires little or no redistribution of charge
among them, and slow or unobserved in the opposite case.
    In Reaction (32), redistribution of charge will be
minimized if A is on the positive end of $BA^+$, so that it retains
as much as possible of the positive charge it had before
reaction.  This will be the case if B is more electronegative
than A, but if B is less electronegative than A, the positive
center in the product ion will shift to B.  This requires that
electrical work be done, and such reactions are predicted to be
very slow or unobserved.
    The correlation of experimental results via electro-
negativity differences is sufficient to lend credence to the
model, although the points show considerable scatter.  Slow or
unobserved reactions are predicted more successfully than fast
reactions.  Requiring little more than electronegativities, this
theory is impressive in the simplicity it brings to one class of
reaction.
    The discussion of energy closes with a brief sketch of
another reaction-rate theory, one native to ICR studies.

ADO Theory.[33]  The Langevin theory was developed for the case of
molecules with no permanent dipole moment.  It is not expected
to be valid for dipolar molecules; this was recognized by
Gioumousis and Stevenson in their derivation of Equation (16).

The effect of the molecular dipole on the potential energy
and the rate constant has been assessed in a "locked dipole"
theory, in which it is assumed that the dipole orients itself
toward the ion at an angle of zero degrees, hence as much as
possible.  This maximization of the dipole contribution leads to
calculated thermal rate constants that are generally unsatis-
factory, being excessively large.[34]

A recent theory takes account of the ion-dipole interaction
by writing Equation (33) for the average potential energy:

$$V(r) = \frac{-\alpha e^2}{2r^4} - \left(\frac{\mu_D e}{r^2}\right) \cos \bar{\theta}(r) \qquad (33)$$

Previously unidentified symbols are the permanent dipole moment
$\mu_D$ and the average angle between the dipole and the ion, $\bar{\theta}$; it
is a function of r.  The first term is the Langevin or induced
dipole contribution.  Inclusion of the $\cos \bar{\theta}(r)$ term for the
average dipole orientation (ADO), which was absent in the locked
dipole theory, reduces the dipolar contribution to V(r).

A kinetic term is added to V(r), converting it to an effective potential energy, $V_{eff}(r)$.  A parameter $r_k$, the critical
distance of approach, is found by equating $\left(\partial V_{eff}(r)/\partial r\right) = 0$.
Distances less than $r_k$ lead to a capture collision.  Two key
parameters can be written using $V_{eff}(r)$; they are the angle
averaged microscopic capture cross section, $<\sigma>$, and the relative velocity, g.  The cross section is a function of the
relative velocity.  A difficult calculation yields the average
dipole orientation, $\bar{\theta}(r)$.  With all of this in hand, the
macroscopic thermal rate constant is given by

$$k_{therm} = \int_0^\infty g<\sigma(g)> \, P(g) \, dg \qquad (34)$$

where P(g) is the Maxwell-Boltzmann velocity distribution in
terms of the relative velocity and the reduced mass.  The
integration is performed numerically.

The theory has been tested with data from proton-transfer
reactions, with quite good results.  A graph of the proton-
transfer reaction between $CH_3OH^+$ and $CH_3OH$ is shown in Figure
4-15.[C42]  The methanol reaction is particularly convenient
because it is the only reaction of the parent ion at thermal
energies.  Rate constants were calculated for the nonthermal

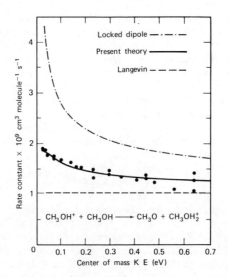

Figure 4-15.  Dependence of total rate constant on
center of mass kinetic energy.  (Reprinted, by permission, from *J. Chem. Phys. 58*, 5175 (1973), Figure 1.)

energies assuming $k = g\langle\sigma\rangle$; the relative velocity was derived
from the energy absorbed by the ion in a radiofrequency field.

Experimental rate constants and three sets of theoretical
rate constants (ADO, Langevin induced dipole, and locked dipole)
are shown in Table 4-1.  The ADO theory is a clear improvement
over its predecessors in explaining the rates of these
reactions.

QUASIEQUILIBRIUM THEORY

The statistical-mechanical treatment of bimolecular reactions
known as absolute rate theory (ART) has been applied to the
unimolecular fragmentation reactions of ions in a mass
spectrometer.[35]  The basic ART equations have been derived in
various ways, most of which depend on the assumption of some
equilibrium or other, whether between the activated complex and
the reactant molecules, or among the vibrational modes of the

Table 4-1.  Rate Constants ( $\times 10^9$ cm$^3$ molecule$^{-1}$ s$^{-1}$) for
Reactions of CH$_5^+$, C$_2$H$_5^+$, C$_3$H$_7^+$ and C$_4$H$_9^+$ with (CH$_3$)$_2$NH

| | | | Theory | | | |
|---|---|---|---|---|---|---|
| Species | Experiment | ADO | Induced dipole | Locked dipole | Experiment/ADO | Theory |
| CH$_5^+$ | 2.15 | 2.00 | 1.62 | 4.31 | 1.08 | |
| C$_2$H$_5^+$ | 1.88 | 1.67 | 1.35 | 3.61 | 1.12 | |
| C$_3$H$_7^+$ | 1.64 | 1.50 | 1.21 | 3.23 | 1.09 | |
| C$_4$H$_9^+$ | 1.09 | 1.40 | 1.13 | 3.02 | 0.78 | |

activated complex.  Because the equilibrium assumption is more
tenuous for ions formed by electron impact than for molecules
involved in an ordinary gas-phase thermal reaction, and because
less is known of the energy microstates of such ions than of
neutral molecules, the term "quasiequilibrium theory" (QET) is
appropriate to the extension of ART to the reactions of interest
here.[36]  It has been applied to simple gaseous ion-molecule
reactions.[C14]
     As ART requires, we write separate statistical-mechanical
expressions for the reactant ion and for the activated complex.
The former becomes the latter when the distribution of its
energy among the available harmonic oscillators causes it to
break apart along the reaction coordinate that leads to the
observed products.  A fundamental question concerns the extent
of randomization of the available energy before the formation of
products.
     Let $\rho^{\ddagger}(\varepsilon)$ be the density of energy states for all degrees
of freedom of the activated complex with the exception of the
reaction coordinate; $\rho(E)dE$ is the number of energy states of
the reactant with energy between E and E + dE, and $E_0$ is the
activation energy for the reaction.  For a molecule or ion with
energy E ( $>E_0$) the reaction-rate constant will be:

$$k(E,E_0) = \frac{1}{h} \int_{\varepsilon = 0}^{E - E_0} \frac{\rho^{\ddagger}(\varepsilon)}{\rho(E)} \, d\varepsilon$$

where h is Planck's constant. The ease of energy transfer
within a system depends on the proximity of the energy states to
each other, hence on the density of states. The rate constant
depends on the density of states in the activated complex--on
the ease with which the energy can reach the reaction coordinate. The integration is over all possible ways of distributing
the energy $E-E_0$ between the reaction coordinate and the rest of
the activated complex.

The theory says nothing and requires nothing about the way
in which the activated complex is formed. The reacting ion-
molecule pair in the Langevin model may be regarded as a suitable basis for such rate-constant calculations. The situation
is a favorable one, because the energies are more or less known
and are not large. A further requirement must be met: the ion-
molecule collision complex must remain intact long enough for
the excess energy to be distributed more or less randomly among
the harmonic oscillators that constitute the complex. The time
required is no more than a few rotation periods. An important
class of ion-molecule reactions has been found that meets this
criterion.[37]

The rate-constant equation is usually written in another
form, more easily handled in practice:

$$k(E,E_0) = \frac{1}{h}\left[\frac{w^{\ddagger}(E - E_0)}{dW(E)/dE}\right]$$

where $W(E)$ is the total number of states with energy less than
or equal to E. The superscript ($\ddagger$) again denotes the activated
complex. There are several approaches to the calculation of
$W(E)$. The one used for studies of ICR reactions assumes that
oscillators of lower frequency will tend to be multiply excited,
while those of higher energy will be unexcited at the energies
involved (0.5-5.0 eV).[38] An expression for $W(E)$ can be written
in terms of the number of oscillators and the geometric mean of
their vibration frequencies.

The process $A^+ + B \rightarrow (AB^+) \rightarrow C^+ + D$ may be viewed as
suggested in Figure 4-16. The decomposition of $(AB^+)$, the
intermediate, is treated according to the QET. The energy E,
assuming formation of $(AB^+)$ from the ground states of $A^+$ and B,
is simply:

$$E = \Delta H_f(A^+) + \Delta H_f(B) - \Delta H_f(AB^+) + \frac{\mu g^2}{2} + E_{int} \qquad (35)$$

where the last two terms are the translational and internal
energies respectively of $(AB^+)$. The term $E_0$ is

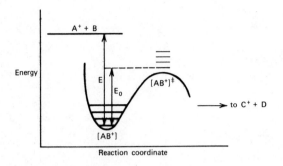

Figure 4-16

$$E_0 = \Delta H_f(AB^+)^{\ddagger} - \Delta H_f(AB^+)$$

Calculation of the rate constant for a specific system requires a large number of data, and these will not all be available even for the simplest of systems. The shortage must be covered by assumptions and estimates. In an example to be considered below, butadiene is chosen as the most plausible isomer for an $AB^+$ system of formula $C_4H_6^+$. The various enthalpies of formation are usually available from the literature, for ions as well as molecules. Vibration frequencies for each oscillator in $AB^+$ and for each one in plausible activated complexes must be assigned. Because exact frequencies are not known for the ion, the corresponding frequencies of the neutral(s) are usually applied in $AB^+$. The vibrations in $(AB^+)^{\ddagger}$ are still more speculative, and are assigned somewhat arbitrarily from a consideration of the probable changes in $AB^+$ as it reacts.

Results of the calculations on the reactions of Table 4-1 were in impressive agreement with the experiments. Where the reactants could form two sets of products, the ratio of products was calculated and compared with experiment. Energies as defined by Equation (35) were approximately 2.5-4 eV for the reactions in Table 4-2.

Another test of the theory was in the isotope effect calculated for the reactions of partially deuterated ions and molecules. A few data are shown in Table 4-3.

The great amount of information required for these

Table 4-2.  Some Product Distributions

| Reaction | | Experiment | Theory |
|---|---|---|---|
| $C_2H_4^{+\cdot} + C_2H_2$ | $\rightarrow C_4H_5^+ + H\cdot$ | 0.19 | 0.18 |
| | $\rightarrow C_3H_3^+ + CH_3\cdot$ | 0.81 | 0.82 |
| $C_2H_2^{+\cdot} + C_2H_4$ | $\rightarrow C_4H_5^+ + H\cdot$ | 0.39 | 0.40 |
| | $\rightarrow C_3H_3^+ + CH_3\cdot$ | 0.61 | 0.60 |
| $C_2D_4^{+\cdot} + C_2D_4$ | $\rightarrow C_4D_7^+ + D\cdot$ | 0.051 | 0.047 |
| | $\rightarrow C_3D_5^+ + CD_3\cdot$ | 0.95 | 0.95 |

calculations makes QET seem impractical for routine use in the
analysis of ion-molecule reactions.  However the application of
the theory to these results is important for the theory, because
it provides a more accurate test of the extent of randomization
of the available energy than do most fragmentation patterns
obtained from mass spectrometry.

Table 4-3.  Some Isotopic Product Distributions

| Reaction | | Experiment | Theory |
|---|---|---|---|
| $CH_2CD_2 + CH_2CD_2^{+\cdot}$ | $\rightarrow C_4H_3D_4^+ + H\cdot$ | 0.68 | 0.71 |
| | $\rightarrow C_4H_4D_3^+ + D\cdot$ | 0.32 | 0.29 |
| $C_2H_2^{+\cdot} + C_2D_2$ | $\rightarrow C_4HD_2^+ + H\cdot$ | 0.59 | 0.65 |
| | $\rightarrow C_4H_2D^+ + D\cdot$ | 0.41 | 0.35 |

A STATISTICAL-MECHANICAL THEORY OF THE RATES OF GASEOUS
BIMOLECULAR REACTIONS

To understand the interaction of theory and experiment in any
field of research is to view that research in its proper
scientific perspective. In some fields the experimental results
seem to be neatly arrayed, awaiting the calculations of the
theoreticians. In others, theory has outrun experiment, so that
the data required to test the theories are chronically in short
supply.
   Chapter 3 attests to ICR spectrometry as a copious source
of data, but the relationship between the data and theories of
gaseous ion-molecule reactions is not that of either of the
above extremes. ICR experiments can provide rate constants,
usually with the added information that a given rate constant
increases or decreases as a function of ion translational
energy. The ionization or appearance potential of the ion is
usually known, as is perhaps the polarizability or dipole moment
of the molecule. That these are, in sum, very meager data
compared to the demands of a detailed reaction theory is clear
from the following brief inspection of a general collision
theory of the rates of bimolecular reactions in the gas phase.[39]
   In this theory the rate of the bimolecular reaction
A + B → C + D is treated as the sum of the rates of all possible
reactions

$$A_i + B_j \longrightarrow C_k + D_l \tag{36}$$

where i, j, k, and l specify for the respective molecules a
complete set of rotational, vibrational, and electronic quantum
numbers; i thus describes the internal state of the molecule $A_i$
prior to its interaction with $B_j$. Translational energy is
treated separately. It is assumed that the rate of Reaction
(36) is a function of i and j, as well as of E, the initial
relative translational energy of $A_i$ and $B_j$. The rate of
Reaction (36) is, in the usual written form,

$$-\left(\frac{dn_{ai}}{dt}\right)_{j,kl} = k_{ij}{}^{kl}(T)\, n_{ai}\, n_{bj}$$

where $k_{ij}{}^{kl}$ is the rate constant corresponding to (36), and the
n values are the number densities of the respective molecules.
The overall rate of reaction of all A and B molecules is:

$$\frac{-dn_a}{dt} = \sum_{ij} \sum_{kl} k_{ij}{}^{kl}\, n_{ai}\, n_{bj} \tag{37}$$

The number of reactive collisions of the molecules in Reaction (36) can be written in terms of a reaction cross section $C_{ij}{}^{kl}(E)$ by means of kinetic theory. The general rate for the reaction A + B becomes, by appropriate substitutions into Equation (37),

$$\frac{-dn_a}{dt} = \sum_{ij} \sum_{kl} \left( \iint f_{ai} f_{bj} C_{ij}{}^{kl}(E) \; g_{ij} \; d\underline{v}_{ai} \; d\underline{v}_{bj} \right) n_{ai} \, n_{bj} \qquad (38)$$

The f values are the unspecified distribution functions for $A_i$ and $B_j$; $g_{ij}$ is the magnitude of the initial relative velocity, and $\underline{v}_{ai}$ and $\underline{v}_{bj}$ are the velocity vectors of the molecules $A_i$ and $B_j$.

Thus far no functional form has been assumed for the distribution functions $f_{ai}$ and $f_{bj}$. Reactants need not be at thermal equilibrium.

The reaction cross sections $C_{ij}{}^{kl}$ of Equation (38) are difficult to obtain, whether by computation or experiment. The quantum-mechanical calculation requires that a solution be found to the time-dependent Schrödinger equation for $A_i$ and $B_j$ as they come together. Depending on the degree of rigor desired, the problem can be partly or wholly solved by classical mechanics in favorable cases.[40]

For the case in which the translational motion of the reactant molecules obeys a Maxwell-Boltzmann distribution, the rate constant for Reaction (36) becomes:

$$k_{ij}{}^{kl}(T) = (\pi\mu)^{-1/2} (2/kT)^{3/2} \int_0^\infty C_{ij}{}^{kl}(E) \; E \; \exp(-E/kT) \; dE$$

When the reaction cross section is in a convenient form, $k_{ij}{}^{kl}$ can be obtained by direct application of a Laplace transform.

The theory thus requires more detailed knowledge of the reacting molecules than ICR affords. This is also true of molecular beam experiments, but to a lesser extent. Beam molecules are often velocity-selected, so that the distribution functions are known. In ICR, the chief problem concerns the electronic and vibrational states of the ion.

NOTES

1.  This is Equation (23.11) in R. P. Feynmann, R. B. Leighton, and M. Sands (eds.), *The Feynmann Lectures on Physics*, Vol. I, Addison-Wesley, Reading, Mass. (1963). Chapter 23 has greatly influenced the present brief treatment and should be read by anyone seeking further information on this subject.

2.  In an ICR spectrometer, $10^{-4}$ torr is a high pressure.

3.  The lifetime and decay modes of $SF_6^-$ are much more complex than this brief statement would imply (see the works by Odom et al.[C68] and Foster et al.[C69]). The latter was discussed on pages 25 and 130.

4.  Huntress[C27] gives it as $A(\infty, \omega_0)$ where $\infty$ is the time (relative to $\xi^{-1}$) and $\omega_0$ is a reminder that the ions are exactly in resonance with $\varepsilon_{rf}$.

5.  The first expression for the force is not familiar to chemists. It is given in a different form in J. O. Hirschfelder, C. F. Curtiss, and R. B. Bird (eds.), *Molecular Theory of Gases and Liquids*, Wiley, New York (1954), p. 852.

6.  *Ann. Chim. Phys.*, Ser. 7 (28), (1903). The derivation of interest is found on pp. 317-318. We have used the current notation in place of Langevin's expression for polarizability. It is sometimes incorrectly implied that the equation is derived in his more famous 1905 paper (Note 12).

7.  For no apparent reason, Langevin, in the passage here translated, gives a positive force at the end of Equation (8), although the derivative requires that it be negative. Moelwyn-Hughes (*Physical Chemistry*, 2nd rev. ed., Pergamon, Oxford (1961), Chapter VII, Equation (20)) quotes Equation (10) and gives an alternate derivation of it.

8.  A table of values is given by Hirschfelder, Curtiss, and Bird, op. cit., p. 989.

9.  An excellent discussion of collision dynamics is that of E. F. Greene and A. Kupperman, *J. Chem. Educ. 45*, 361 (1968).

10. When a reaction cross section is known as a function of the relative velocity or relative translational energy of the collision partners, it is referred to as an "excitation function." (See R. L. LeRoy, *J. Phys. Chem. 73*, 4338 (1969).)

11. For a critical review of both theoretical and experimental problems, see M. Henchman, Rate constants and cross sections, in J. L. Franklin (ed.), *Ion-Molecule Reactions*, Plenum, New York (1972), Vol. 1.

12. This was Langevin's choice (P. Langevin, *Ann. Chim. Phys.* 5, 276 (1905)). Today we usually assume the molecule to be at rest.

13. The symbol $\sigma$ is sometimes used to denote $r_A + r_B$, in which case the collision cross section is $\pi\sigma^2$.

14. P. Langevin, *Ann. Chim. Phys.* 5, 245 (1905). Part of this paper has been translated by E. W. McDaniel, *Collision Phenomena in Ionized Gases*, Wiley, New York (1964), Appendix II. See also G. Gioumousis and D. P. Stevenson, *J. Chem. Phys.* 29, 294 (1958), and *Ion-Molecule Reactions in the Gas Phase*, Advances in Chemistry Series, No. 58, American Chemical Society (1966).

15. Collisions are here defined by Equation (14), not Equation (12). For a discussion of the activation-energy question, see the work by Henis.[E22]

16. See, for example, R. Wolfgang, *Acc. Chem. Res.* 2, 248 (1969), and C. E. Melton and G. A. Neece, *J. Am. Chem. Soc.* 93, 6757 (1971).

17. They move toward each other. Our frame of reference is that of an observer on the ion.

18. This k is sometimes called the "capture reaction-rate constant" in the recent literature, the point being that this is the rate at which ions "capture" molecules. If more than one set of products is observed in the reaction of a particular ion and molecule, this k will be equal to the sum of the rate constants for the various reactions.

19. G. Gioumousis and D. P. Stevenson, *J. Chem. Phys.* 29, 294 (1958); D. P. Stevenson, Ion-molecule reactions, in C. A. McDowell (ed.), *Mass Spectrometry*, McGraw-Hill, New York (1963).

20. Compare the works by Bowers et al.[C6] and Clow et al.[C20]

21. For further discussion see the work by Dunbar,[C2] Chapters 7 and 8 in *Ion-Molecule Reactions in the Gas Phase*, Advances in Chemistry Series, No. 58, American Chemical Society (1966), and especially two short papers by Mackenzie Peers: *Int. J. Mass Spectrom. Ion Phys.* 3, 99-102 (1969), in which a novel derivation of Equation (16) is given, and *Int. J. Mass Spectrom. Ion Phys.* 4, 251-252 (1970).

22. Here the $\tau$, $\tau'$ notation used by Bowers et al.[E9] has been reversed in order to bring it into conformity with later usage in this book and elsewhere.

23. The calculation from linewidths requires careful attention to units.

24. The work by McAllister[C32] has a misplaced factor of 2.

25. The literature employs a confusing variety of symbols for the times during which the ions are in the source and the analyzer. This t' is with reference to t' = 0 at the

electron beam.  In order for Equation (27) to be correct in the analyzer as well as in the source, t' must be defined and continuous in both regions.  Because the ion leaves the source at $\tau'$, and because it is convenient in integration of the power absorption to have a time variable that is zero at the entrance to the analyzer, t' is replaced in the analyzer by $\tau' + t$.

26.  Residence time and drift time are synonymous.

27.  Reference C61 erroneously gives the $H_2O$ pressure as $1.06 \times 10^{-9}$ torr.

28.  See, for example, R. T. Weidner and R. L. Sells, *Elementary Modern Physics*, 2nd ed., Allyn and Bacon, Boston (1968), pp. 437-439.  Their $E_{CM}$ is the energy *unavailable* for reaction because momentum must be conserved.

29.  The ionization potential of ammonia is 10.154 V.

30.  M. Henchman, Rate constants and cross sections, in J. L. Franklin (ed.), *Ion-Molecule Reactions*, Plenum, New York (1972), Vol. 1.

31.  Ion translational energies are near their thermal values in the absence of irradiation from a radiofrequency oscillator.

32.  R. W. Kiser, *Introduction to Mass Spectrometry and Its Applications*, Prentice-Hall, Englewood Cliffs, N.J. (1965), Chapter 8.

33.  See works by Su et al.[C41,C51-C53,C55] and Bowers et al.[B45,C42]

34.  S. K. Gupta, E. G. Jones, A. G. Harrison, and J. J. Myher, *Can. J. Chem. 45*, 3107 (1967).

35.  H. M. Rosenstock, M. B. Wallenstein, A. L. Wahrhaftig, and H. Eyring, *Proc. Nat. Acad. Sci. U.S. 38*, 667 (1952).

36.  R. W. Kiser, *Introduction to Mass Spectrometry and Its Applications*, Prentice-Hall, Englewood Cliffs, N.J. (1965), Chapter 7 and references therein.

37.  R. Wolfgang, *Acc. Chem. Res. 3*, 48 (1970).

38.  This hypothesis was first worked out by Vestal et al.

39.  M. A. Eliason and J. O. Hirschfelder, *J. Chem. Phys. 30*, 1426 (1959).  However a number of ICR experiments have been done on small molecules in order to test simple reaction theories.

40.  See E. F. Greene and A. Kuppermann, *J. Chem. Educ. 45*, 361 (1968).

CHAPTER 5 Photon Studies and
Other Special Experiments

PHOTON STUDIES

Primary ions are usually produced in an ICR spectrometer by
direct ionization at the electron beam, and secondary ions by
chemical reaction in the dark. The introduction of light into
the cell can produce ions by several processes that were
characterized before the advent of ICR. There is *photoionization* of an atom or molecule,

$$P \xrightarrow{h\nu} P^+ + e^-$$ (1)

and *photodetachment* of an electron from an anion,

$$P^- \xrightarrow{h\nu} P + e^-$$ (2)

These two are so similar that it is difficult to justify the use
of different terms for them. A somewhat different reaction is
*photodissociation*,

$$P^+ \xrightarrow{h\nu} S^+ + neutral$$ (3)

in which the reactant ion[1] dissociates as it absorbs a photon.
Since each of these processes involves at least one ion, their
study by ICR spectrometry is straightforward in principle,
although it presented some experimental problems at its
inception.[E29]
    The quantitative study of such reactions is a rich source
of chemical information. It provides data for calculation of
the electronic states of ions and puts theories of unimolecular
dissociation to the test. When energy thresholds can be
established Reaction (2) gives electron affinities and Reaction
(3), bond-dissociation energies.
    To maximize the extent of the photoreactions, the ions are

held in the cell longer than usual.  The results of a photodetachment study of $SiH_3^-$ and $SiD_3^-$ are shown in Figure 5-1.[F11]
Typical pressures were in the vicinity of 1 μtorr.  The anions
are generated in the reaction

$$F^- + SiH_4 \longrightarrow HF + SiH_3^-$$

a "clean," direct source of $SiH_3^-$.  The $F^-$ is produced by
dissociative electron capture of $NF_3$ at 1.7 eV.  A 1000 W xenon
lamp provides the light for the photodetachment.

   An extensive data analysis yields the wavelength-dependent
relative photodetachment cross sections.[F11]  The $SiH_3^-$ experiments lead to a value for the electron affinity of $SiH_3 \cdot$ that is
no greater than 1.44 ± 0.03 eV.

   A novel apparatus in which the ICR cell is set into an
extended laser cavity yields a substantial increase in the
photodetachment effect.[A42]  The ICR cell and laser are shown in
Figure 5-2 and the results of two photodetachment experiments on
$OH^-$, in Figure 5-3.  On the left the laser and the ICR cell are
separate, and the light is reflected back through the cell by a
mirror (a double-pass experiment).  On the right is the signal
when the ICR cell is incorporated into the laser (an intracavity

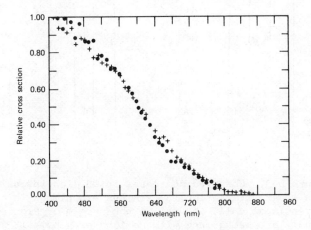

Figure 5-1.  Relative photodetachment cross section for
$SiH_3^-$ (+) and $SiD_3^-$ (●).  (Reprinted, by permission,
from *J. Chem. Phys. 61*, 4833 (1974), Figure 1.)

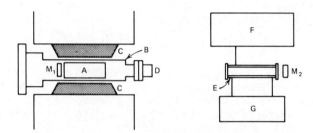

Figure 5-2.  Schematic view of the optical system used
for the intracavity experiments.  $M_1$ and $M_2$ are
mirrors, A is the ICR cell, B a vacuum chamber, C the
pole faces of the electromagnet, D the viewing port
window, E the coaxial dye-cell flashlamp, F the capacitor and power supply, and G the dye-circulation
system.  (Reprinted, by permission, from *Rev. Sci.
Instrum. 45*, 1154 (1974), Figure 1.)

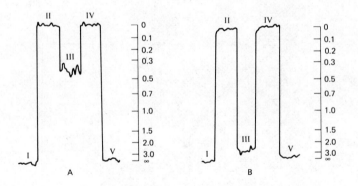

Figure 5-3.  Observed decreases in $OH^-$ ion intensity
upon laser irradiation in:  (A) an experiment in a
conventional cell, and (B) in the system of Figure 5-2.
I and V are background with no ions present.  II and IV
are the maximum ion intensity with no laser irradiation.  III is the ion intensity 2.5 ms after laser
irradiation.  (Reprinted, by permission, from *Rev. Sci.
Instrum. 45*, 1154 (1974), Figure 2.)

experiment) as in Figure 5-2. The Rhodamine 6G dye laser is
operated at a wavelength of 599 nm and produces about 8 mJ per
pulse. The signals are the sum of not less than 10 ion-
formation and laser-pulse cycles.

Use of laser light and an ICR cell for photodetachment
studies is very promising. The ions can be formed by electron
impact or by chemical reaction, and the high intensity and
generally narrow bandwidth of the laser light facilitate precise
studies at low concentrations, or of species with small photodetachment cross sections. If the cell is operated in the ion-
trapping mode, the laser pulse can be delayed until the ions
have attained nearly thermal energies by collision, or until an
interfering ion is eliminated by reaction.[A41]

Photodetachment is the expected result when anions are
illuminated, but photodissociation is the predominant process in
the case of $Fe(CO)_4^-$ and $Fe(CO)_3^-$. The reaction system is[F10]

$$Fe(CO)_4^- \xrightarrow{h\nu} Fe(CO)_3^- \xrightarrow{h\nu} Fe(CO)_2^-$$

$Fe(CO)_3^-$ is more reactive than $Fe(CO)_4^-$, as revealed by its
reactions with $Fe(CO)_5$ and $SF_6$. Data for the photodissociation
of $Fe(CO)_4^-$ are summarized in Figure 5-4. Experiments were
performed in a drift cell with high trapping voltages and a
$Fe(CO)_5$ pressure near $10^{-6}$ torr. As a check on the validity of
the experiment it was shown by ejection of $Fe(CO)_4^-$ that it does
not produce $Fe(CO)_3^-$ in the dark.

The previous systems presented no serious experimental
complications. In contrast, the observation of the photodissociation of chloromethane cations

$$CH_3Cl^+ \xrightarrow{h\nu} CH_3^+ + Cl$$

in a conventional drift cell is more difficult.[E29] The product
ion is also a primary ion formed at the electron beam. Moreover
it is necessary to show that it is not being formed in an ion-
molecule reaction; this was done in an ion-ejection experiment.
A conventional light source was used, and the drift voltages
were chosen so as to hold the ions in the cell for about 1 s,
thus allowing ample time for photon-ion interaction. With such
a long ion residence time, it is more realistic to deal with
concentrations in a "leaky" cell than with ion currents. The
rate constant $k_\lambda$ for the photodissociation can be obtained as
follows. Let $(P)$ be the concentration of $P^+$ ions under
illumination. $(S)$ and $(S_0)$ are the concentrations of $S^+$ ions
with and without illumination respectively. $k_S^{loss}$ is the rate
constant for loss of $S^+$ from the cell. By definition it is

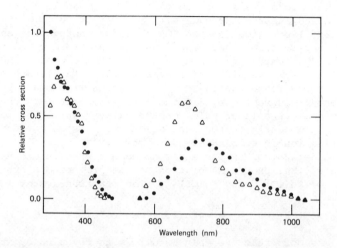

Figure 5-4.  Relative photodissociation cross sections.
($\bullet$) $Fe(CO)_4^-$ disappearance; ($\Delta$) $Fe(CO)_3^-$ appearance.
(Reprinted, by permission, from *J. Amer. Chem. Soc. 96*,
3671 (1974), Figure 1.)

independent of the number of $S^+$ ions, and thus independent of
the photodissociation reaction.  $C_S$ is the ion current of $S^+$
formed at the electron beam.

The kinetic equation for the production and disappearance
of $S^+$ is

$$\frac{d(S)}{dt} = C_S - k_S^{loss}(S) + k_\lambda(P) \tag{4}$$

Two steady states exist.  In the dark,

$$\frac{d(S_0)}{dt} = 0 = C_S - k_S^{loss}(S_0)$$

so that

$$C_S = k_S^{loss}(S_0) \tag{5}$$

The steady state under illumination is

$$\frac{d(S)}{dt} = 0 = C_S - k_S^{loss}(S) + k_\lambda(P)$$

Solving this expression for $k_\lambda$ and using Equation (5) to eliminate $C_S$ gives

$$k_\lambda = k_S^{loss} \frac{(S) - (S_0)}{(P)} \tag{6}$$

This equation gives the photodissociation rate constant in terms of measurable parameters, although $k_S^{loss}$ is obtained indirectly and must be regarded as somewhat uncertain.

The photodissociation rate constant depends on the monochromatic light intensity $\phi(\lambda)$ and on the photodissociation cross section $\sigma(\lambda, P^+)$. The latter depends on the nature of the ion as well as on the wavelength of the light.

$$k_\lambda = \phi(\lambda)\sigma(\lambda, P^+) \ s^{-1} \tag{7}$$

Equations (6) and (7) can be solved for the cross section, for which absolute values are obtained if $\phi(\lambda)$ is known (see Figure 5-5). The threshold in Figure 5-5 leads to the uncertain value of 2.8 eV for the dissociation threshold of $CH_3Cl^+$. The gentle slope of the cross section at low energy suggests that a more sensitive means of detection might reveal dissociation at lower energies. The light intensity was measured as a function of wavelength; this variation is considered in constructing the curve of Figure 5-5.


ION EJECTION

Not infrequently in the development of a scientific technique, what was first seen as a bothersome aspect of the experiment is later turned to advantage, and leads to a refinement of the measurements or an extension of the capabilities of the instrument. So it was with ion ejection. The commercial instrument includes an ion-current meter, and users of these spectrometers could not have failed to note occasional variations in the ion current in the course of a scan. However the meaning of this information was not immediately appreciated.

Throughout Chapter 1, and especially in discussing ion dynamics, we assumed that the ions traveled through the source and analyzer regions, and that they came into contact only with the ion-current collector plates at the end of the cell, beyond the region in which their movement was subject to drift and

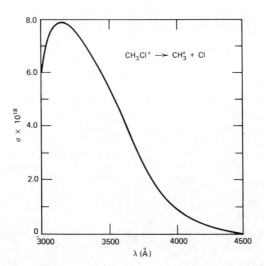

Figure 5-5. Photodissociation of $CH_3Cl^+$ to $CH_3^+$ and
Cl·. The absolute cross section for reaction is shown
as a function of wavelength. (Reprinted, by permission, from *J. Amer. Chem. Soc. 93*, 4354 (1971),
Figure 4.)

trapping fields. In Chapter 2 the thermodynamic interpretation
of double-resonance signals was subject to a caveat: a signal
indicating a decrease in the number of product ions could be due
to a decrease in the number of reactant ions if the latter
absorbed enough energy from the irradiating oscillator to reach
the upper and lower plates of the source region. Double-
resonance signals are unambiguous only when obtained under the
gentlest of conditions. In practice this means that the
operator must choose as low an amplitude as possible for the
irradiating oscillator and resist the temptation to boost the
signal intensity by working at high emission currents.
    The technique of ion ejection uses an irradiating oscilla-
tor to eliminate the ions of a certain m/e from the cell. The
amplitude of the radiofrequency radiation is increased to the
point that it sweeps these ions against the cell plates, where
they are neutralized. Ion ejection is sometimes done in a cell

of conventional design, but a special one with an irradiation
region between the source and analyzer regions has also been
used.[C23]
     The ion parameters that characterize the removal of ions
from the cell can be obtained from the equations of Chapter 1.
The time required to eject an ion, $t_{ej}$, is the time at which the
radius of the ion path is equal to half the distance between
upper and lower plates of the cell: $r = d/2$. The time and the
radius are related by

$$r = \frac{\varepsilon_{rf} t}{2B} \qquad \left(\text{Chapter 1, Equation (10)}\right)$$

which, when solved for t under these conditions is

$$t_{ej} = \frac{Bd}{\varepsilon_{rf}}$$

When it is ejected, the ion has drifted a distance x, which is
related to the ejection time by noting that $(x/t_{ej}) = v_{dr} =$
$(\varepsilon/B)$:

$$x = \frac{\varepsilon t_{ej}}{B} = \frac{\varepsilon}{B} \times \frac{Bd}{\varepsilon_{rf}} = \frac{\varepsilon d}{\varepsilon_{rf}} = \frac{V_{dr}}{\varepsilon_{rf}} \qquad (8)$$

In these equations $\varepsilon_{rf}$ is the maximum intensity of the irradiating rf field, $\varepsilon$ is the intensity of the drift field, $V_{dr}$ is
the drift voltage, equal to $\varepsilon d$, and $v_{dr}$ is the drift velocity.
Equation (8) enables us to calculate the minimum length of the
ejection region of the cell.
     The kinetic energy of an ion is given by

$$E_k = \frac{mv^2}{2} = \frac{m}{2}(\omega r)^2$$

in which v is the velocity of the ion. Writing $\omega = (eB/m)$ and
$r = (d/2)$ yields

$$E_k = \frac{(eBd)^2}{8m}$$

for the kinetic energy of an ion as it strikes the drift plates.
     Selective removal of ions of a certain m/e from the cell is
most useful in the case of a secondary or higher-order ion
produced by more than one reaction. Omitting the neutral
molecules, such a reaction system is

A **conventional** double-resonance experiment will show that $S^+$ is **produced by** both $P_1^+$ and $P_2^+$ but will not indicate the extent of **product** derived from either $P_1^+$ or $P_2^+$. By removing each of the **reactant** ions in turn, the ion-ejection experiment provides such **information.** Consider the reactions

$$NH_3^+ + NH_3 \xrightarrow{k_2} NH_4^+ + NH_2 \qquad (9)$$
$$m/e \quad 17 \qquad\qquad\qquad 18$$

$$NH_2^+ + NH_3 \xrightarrow{k_3} NH_4^+ + NH \qquad (10)$$
$$m/e \quad 16 \qquad\qquad\qquad 18$$

The relative contributions of the reactant ions $NH_3^+$ and $NH_2^+$ to the common product can be seen in Figure 5-6.[C23] At the left and right of the scan no ions are ejected. This signal height corresponds to the full intensity of the $NH_4^+$ ion. In between are seen the intensities of $NH_4^+$ as first $NH_3^+$ and then $NH_2^+$ are ejected. The intensity with $NH_3^+$ ejected is the fraction of the total intensity due to $NH_2^+$ ions. Likewise the intensity with $NH_2^+$ removed is the contribution of $NH_3^+$ to the formation of product ions. The sum of these two is approximately equal to the full intensity, as it should be if the reaction system that produces $NH_4^+$ is as simple as we have assumed.

One other factor must be taken into account in order to calculate the ratio of rate constants for Reactions (10) and (9). The number of ions of $NH_3^+$ and of $NH_2^+$ prior to reaction are not equal. As these are both primary ions, their relative rates of formation may be obtained from a conventional scan at a pressure sufficiently low that no reactions occur. The result for this system is $k_3/k_2 = 0.6 \pm 0.1$.

A second type of ion-ejection experiment takes advantage of the potential well created by the trapping plates. These plates are at the same positive potential, and create a nearly parabolic electric field with a minimum at the middle of the cell. This is exactly what is needed for effective trapping. Ions oscillate in this potential well at an angular frequency of[A6]

$$\omega_T = \left(\frac{4V_T e}{md^2}\right)^{1/2} \qquad (11)$$

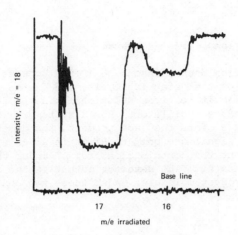

Figure 5-6.  Ion-ejection spectrum taken by monitoring
the intensity of $NH_4^+$ while sweeping the irradiating
oscillator at high power over the resonant frequencies
of $NH_2^+$ and $NH_3^+$.  The radiofrequency power is switched
into the cell at the point where the beat pattern
between the two oscillators begins.  (Reprinted, by
permission, from *J. Chem. Phys. 54*, 843 (1971), Figure
5.)

where $V_T$ is the trapping voltage and d the distance between the
trapping plates (the width of the cell).  This oscillation can
be pumped to the point that ions touch the trapping plates and
are neutralized.  It is done by applying an oscillating field of
frequency $\omega_T/2\pi$ to these plates.  The technique suffers from
poor mass resolution ($m/\Delta m \simeq 4$) that practically limits it to
studies of light ions or simple reaction systems involving ions
of very different m/e ratios.  It is nonetheless interesting
because it is based on an aspect of ion dynamics that is usually
ignored.

Ion ejection via the trapping plates has also been done in
a four-section cell that features two pairs of trapping plates,
one for the source and the other for the reaction and analyzer
regions.[E39]  The radiofrequency ejection wave can be applied to
either pair of plates.  This cell was used for the chemical

ionization studies of the $C_6$ hydrocarbons as discussed in
Chapter 2.  There is some problem with mass resolution, but it
does not appear to be as severe as the figure quoted above for
the standard cell.

### Pulsed Ion-Ejection Double Resonance

Pulsed ion-ejection double resonance (PIEDR) ejects one reactant
ion in one part of the detection cycle, and in the other part
this ion reacts as it acquires additional kinetic energy from
the double-resonance oscillator.  For a system in which two
reactants $A^+$ and $B^+$ form the same product, the detector receives
at one moment a signal due only to $B^+$ while $A^+$ is ejected, but
at the next moment the contribution from $A^+$ under the influence
of $\varepsilon_{rf}$ is added to it.  The detector output is the difference in
the number of product molecules under these two conditions.[A19]

## TOTAL ION-CURRENT STUDIES

The removal of ions of a particular m/e has been observed
directly by measurement of the total ion current collected at
the end of the cell.[C21]  If the electric field intensity of the
irradiating oscillator is sufficiently high, resonant ions will
reach the top and bottom plates of the cell rather than the ion-
current plates; their loss will appear as a decrease in the
total ion current when the resonant condition is met.  The
marginal oscillator is not needed.

A total ion-current scan can be done by variation of either
the magnetic field intensity or the oscillator frequency.  In
the latter case, all ions drift through the cell at the same
speed (Chapter 1, Equation (3)).  This is convenient when
studying reaction rates.  If the electric field intensity of the
radiofrequency oscillator is sufficient for removal of all
resonant ions, the signals in a total ion-current scan do not
depend on the mass of the ion, as they do in conventional
detection by the marginal oscillator (Chapter 1, Equation (7)).
A total ion-current scan of $CH_2FCl$ is shown in Figure 5-7.

Double-resonance experiments are straightforward in total
ion-current measurements.  A reactant ion can be irradiated at
varying electric field intensities in the source and the product
ion current observed in the analyzer.  The resulting change in
the product ion current will be due to the change in the rate
constant as a function of the kinetic energy of the reactant
ion, as long as that ion is not swept against the cell plates.

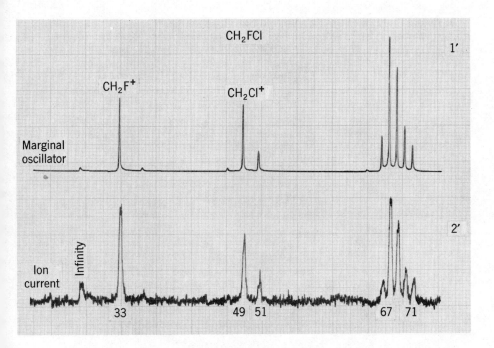

Figure 5-7.  Marginal oscillator (upper) and total ion
current (lower) scans of $CH_2FCl$.  (Figure courtesy of
K. R. Jennings.)


*Pulsed Ion Cyclotron Double Resonance*

The earliest example of a time separation experiment is pulsed
ion cyclotron double resonance,[C12] first done in a drift cell.
The schematic description in Figure 5-8 sets the pattern for
other modifications of this type.  Ions are formed for 0.1 ms,
and immediately thereafter the irradiating oscillator is turned
on for 0.1 ms at the resonance frequency of a known reactant
ion.  This time is short compared to the residence time in the
cell, so that almost all of the ions will absorb the same amount
of energy[2] before reacting.  The energy is given by Equation (9)
of Chapter 1, which shows that it varies with the square of the
electric field intensity $\varepsilon_{rf}$.  Measurement of the product-ion
signal in a series of experiments as a function of $\varepsilon_{rf}$ gives the
dependence of the rate constant on reactant-ion kinetic energy.

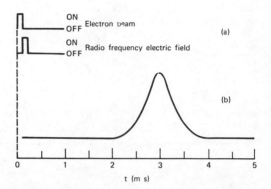

Figure 5-8.  (A) Electron beam and radiofrequency electric field pulse sequence for the production and irradiation of reactant ions.  (B) A typical $C_2H_5^+$ signal obtained after the pulse sequence shown above. (Reprinted, by permission, from *J. Phys. Chem. 73*, 469 (1969), Figures 1(a) and (b).)

The technique has been applied to the study of the two reactions

$$CH_3^+ + CH_4 \longrightarrow C_2H_5^+ + H_2$$

$$CH_3^+ + CH_4 \longrightarrow C_2H_3^+ + 2H_2$$

The first of these is known from thermochemistry to be exothermic, and the rate of reaction decreases with increasing ion energy.  The second reaction is endothermic; the reaction threshold at about 1 eV is in qualitative agreement with the predictions of thermochemistry, and the reaction rate goes through a maximum.

The lower trace in Figure 5-8 shows a $C_2H_5^+$ signal after averaging 10 000 times.  Its onset at about 1.5 ms and disappearance starting at about 3.0 ms are in agreement with the times calculated from drift velocity for entry of the ions into the analyzer region of the cell and their subsequent departure.

PENNING IONIZATION

If an atom or molecule in an excited electronic state returns to
the ground state during a collision in which it ionizes another
atom or molecule, the process is known as "Penning ionization":

$$M^* + N \longrightarrow M + N^+ + e^-$$

In order to allow adequate time for collision, the exciting
species must be in a metastable state (one from which the
transition to the ground state is forbidden), so that it stores
energy for considerably longer than the approximate $10^{-8}$ s
required for an allowed transition to the ground state. Because
the energy of the metastable state $M^*$ must be no less than the
ionization energy of N, the excited species must itself have a
still higher ionization energy. Nitrogen and argon are frequently used as the energy carriers.

The adaptation of Penning ionization to ICR spectrometry is
as shown in Figure 5-9.[C11] A multichannel beam source sends a
collimated beam of molecules into an excitation chamber where
they intersect an electron beam that excites a small fraction of
them. The molecular beam, which now contains some metastable
molecules, crosses a region of positive electric field that
repels any ions formed by the electron beam. The molecular beam
then enters the analyzer region of an ICR cell, into which
molecules of a second species have been admitted. The second
molecule is chosen so that its ionization potential will allow
Penning ionization, and the ions thus formed are detected by the
marginal oscillator.

The Penning ionization of benzene by $N_2$ is shown in Figure
5-10. We see that the ordinate is the $C_6H_6^+$ signal intensity
divided by the electron-emission current. This is done in order
to allow for the variation of emission current with electron
energy. The onset of the curve is about 9.25 eV, the ionization
energy of benzene.

The ICR analyzer cell, which detects ions in the space
where they are formed, is well-suited to Penning ionization
studies.

INELASTIC ELECTRON-MOLECULE COLLISIONS

To conclude this chapter we consider a measurement that can be
made without any instrumental modification.[3] The only change is
in the mind of the user of an ICR spectrometer, who must think
of his instrument in a new way.

The cross section for the dissociative attachment reaction

of ethyl nitrate,

$$e^- + C_2H_5ONO_2 \longrightarrow NO_2^- + C_2H_5O \cdot \qquad (12)$$

is very large for thermal-energy electrons. $C_2H_5ONO_2$ thus serves as a scavenger for such slow electrons, and of necessity generates an ion in the process.

The electrons are constrained by the trapping voltage to oscillate back and forth in the cell, which greatly increases the likelihood of Reaction (12).

A target gas whose excitation potentials are to be studied is introduced into the cell along with the ethyl nitrate, and the energy of the electron beam is varied. When it corresponds to the energy difference between an occupied molecular electronic state and a higher unoccupied one, the impacting electron

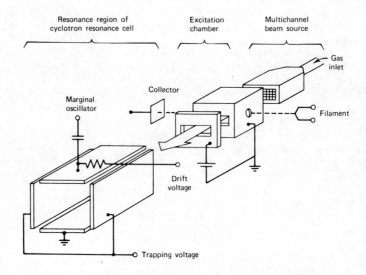

Figure 5-9. Schematic diagram of the modified ICR apparatus for the study of ionizing reactions of metastable atoms and molecules. The electron beam is collinear with the applied magnetic field. (Reprinted, by permission, from *Int. J. Mass Spectrom. Ion Phys.* 3, 149 (1969), Figure 1.)

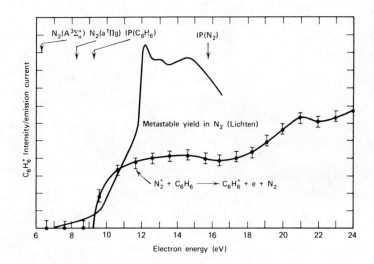

Figure 5-10.  Penning ionization yield of $C_6H_6^+$ by
collision with metastable $N_2$.  (Reprinted, by permission, from *Int. J. Mass Spectrom. Ion Phys. 3*, 149
(1969), Figure 2.)

excites the target molecule to the higher state, and recoils
from its inelastic collision with very little energy.  Reaction
(12) follows, and the $NO_2^-$ ions thus generated are observed in
the usual manner.

The experiment was initially conducted with carbon tetra-
chloride as the scavenger; it dissociates to produce $Cl^-$.[A5]
The same anion is also produced directly at the electron beam, a
complication that is absent when ethyl nitrate is used.  The
inelastic excitation spectrum of $CS_2$ is shown in Figure 5-11.[B5]

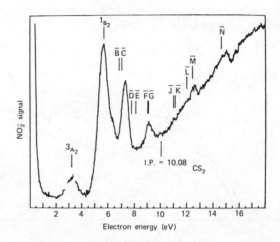

Figure 5-11. Low-energy electron inelastic excitation spectrum of carbon disulfide. Ethyl nitrate was used to detect thermal energy electrons by monitoring the $NO_2^-$ ion signal. (Reprinted, by permission, from *J. Amer. Chem. Soc. 92*, 1128 (1970), Figure 2.)

NOTES

1. The analogous reaction by neutral species is not directly accessible to ICR, as there are no ions involved.
2. We ignore the spread in energies caused by variations in the phase angle between the radiofrequency field and the ions in cyclotron motion.
3. The study of anions requires a negative trapping potential.

# BIBLIOGRAPHY

INSTRUMENTAL DESIGN AND MODIFICATION

A1. L. R. Anders, J. L. Beauchamp, R. C. Dunbar, and J. D. Baldeschwieler, Ion-cyclotron double resonance, *J. Chem. Phys. 45*, 1062 (1966).

A2. J. D. Baldeschwieler, What is ion cyclotron resonance?, *Kagaku no Ryoiki 21*, 785 (1967).

A3. J. M. S. Henis and W. Frasure, Ion cyclotron resonance detection using electron energy modulation, *Rev. Sci. Instrum. 39*, 1772 (1968).

A4. J. D. Baldeschwieler, Ion cyclotron resonance spectroscopy, *Science 159*, 263 (1968).

A5. D. P. Ridge and J. L. Beauchamp, Application of ion cyclotron resonance to the study of inelastic excitation by low-energy electrons, *J. Chem. Phys. 51*, 470 (1969).

A6. J. L. Beauchamp and J. T. Armstrong, An ion ejection technique for the study of ion-molecule reactions with ion cyclotron resonance spectroscopy, *Rev. Sci. Instrum. 40*, 123 (1969).

A7. R. M. O'Malley and K. R. Jennings, Detection of inelastic scattering of electrons by simple polyatomic molecules using the $SF_6^-$ ion, *Int. J. Mass Spectrom. Ion Phys. 2*, App. 1 (1969).

A8. R. T. McIver, Jr., A pulsed grid modulation scheme for an ion cyclotron resonance spectrometer, *Rev. Sci. Instrum. 41*, 126 (1970).

A9. R. T. McIver, Jr., A trapped ion analyzer cell for ion cyclotron resonance spectroscopy, *Rev. Sci. Instrum. 41*, 555 (1970).

A10. G. C. Goode, R. M. O'Malley, A. J. Ferrer-Correia, and K. R. Jennings, Ion cyclotron resonance mass spectrometry, *Nature 227*, 1093 (1970).

A11. G. C. Goode, R. M. O'Malley, A. J. Ferrer-Correia, and K. R. Jennings, Ion cyclotron resonance mass spectrometry, *Chem. Brit. 7*, 12 (1971).

A12.   J. D. Baldeschwieler and S. S. Woodgate, Ion cyclotron resonance spectroscopy, *Acc. Chem. Res. 4*, 114 (1971).

A13.   G. A. Gray, Ion cyclotron resonance, *Adv. Chem. Phys. 19*, 141 (1971).

A14.   J. I. Brauman, Ion cyclotron resonance, *XXIII International Congress on Pure and Applied Chemistry*, Volume 7, Butterworths, London (1971).

A15.   G. C. Goode, R. M. O'Malley, A. J. Ferrer-Correia, and K. R. Jennings, Ion cyclotron resonance mass spectrometry, *Adv. Mass Spectrom. 5*, 195 (1971).

A16.   G. C. Goode, K. R. Jennings, and C. J. Drewery, Ion cyclotron resonance mass spectrometry, Chapter in A. Maccoll (ed.), *Mass Spectrometry*, Vol. V, Physical Chemistry Series 1 (1971).

A17.   R. C. Dunbar, Transient ion cyclotron resonance method for studying ion-molecule collision and charge transfer rates: $N_2^+$ and $CH_4^+$, *J. Chem. Phys. 54*, 711 (1971).

A18.   K. C. Smyth, R. T. McIver, Jr., J. I. Brauman, and R. W. Wallace, Photodetachment of negative ions using a continuously tunable laser and an ion cyclotron resonance spectrometer, *J. Chem. Phys. 54*, 2758 (1971).

A19.   R. T. McIver, Jr., and R. C. Dunbar, Pulsed ion cyclotron double resonance for the study of ion molecule reactions, *Int. J. Mass Spectrom. Ion Phys. 7*, 471 (1971).

A20.   T. B. McMahon and J. L. Beauchamp, Determination of ion transit times in an ion cyclotron resonance spectrometer, *Rev. Sci. Instrum. 42*, 1632 (1971).

A21.   M. L. Gross and C. L. Wilkins, Computer-assisted ion cyclotron resonance appearance potential measurements for $C_5H_{10}$ isomers, *Anal. Chem. 43*, 1624 (1971).

A22.   M. L. Gross and C. L. Wilkins, Ion cyclotron resonance spectrometry: recent advances of analytical interest, *Anal. Chem. 43* (14), 65A (1971).

A23.   J. H. Futrell, Ion cyclotron resonance spectrometry, in D. Price (ed.), *Dynamic Mass Spectrometry*, Vol. 2, Heyden, London (1971), p. 97.

A24.   T. B. McMahon and J. L. Beauchamp, A versatile trapped ion cell for ion cyclotron resonance spectroscopy, *Rev. Sci. Instrum. 43*, 509 (1972).

A25.   T. McAllister, A direct method for obtaining electron impact spectra from ICR mass spectrometer, *Chem. Phys. Lett. 13*, 602 (1972).

A26.   T. McAllister, Electron impact excitation spectra in an ion cyclotron resonance mass spectrometer, *J. Chem. Phys. 57*, 3353 (1972).

A27.  J. M. S. Henis, Ion cyclotron resonance spectrometry, in
      J. L. Franklin (ed.), *Ion-Molecule Reactions*, Vol. 2,
      Plenum, New York (1972), p. 395.

A28.  C. J. Drewery, G. C. Goode, and K. R. Jennings, Ion cyclo-
      tron resonance spectrometry, in A. Maccoll (ed.), *Mass
      Spectrometry*, M. T. P. International Review of Science,
      Physical Chemistry, Butterworths, London (1972), Chapter
      6.

A29.  R. T. McIver, Jr., A solid-state marginal oscillator for
      pulsed ion cyclotron resonance spectroscopy, *Rev. Sci.
      Instrum. 44*, 1071 (1973).

A30.  J. I. Brauman and L. K. Blair, Ion cyclotron resonance
      spectroscopy, in F. C. Nachod and J. J. Zuckerman (eds.),
      *Determination of Organic Structures by Physical Methods*,
      Vol. 5, Academic, New York (1973).

A31.  W. T. Huntress, Jr., and W. T. Simms, A new ion and
      electron detector for ion cyclotron resonance spectros-
      copy, *Rev. Sci. Instrum. 44*, 1274 (1973).

A32.  I. B. Woods, M. Riggin, T. F. Knott, and M. Bloom,
      Dependence of ion cyclotron resonance on cell parameters,
      *Int. J. Mass Spectrom. Ion Phys. 12*, 341 (1973).

A33.  H. Hartmann, K. H. Lebert, and K. P. Wanczek, Ion cyclo-
      tron resonance spectroscopy, *Fortschr. Chem. Forsch. 43*,
      57 (1973).

A34.  J. Sohma, Studies of chemical reactions by the magnetic
      resonance, *Bussei Kenkyu 18*, A41 (1973).

A35.  J. L. Beauchamp, Ion cyclotron resonance spectroscopy,
      *Ann. Rev. Phys. Chem. 22*, 527 (1971).

A36.  M. B. Comisarow and A. G. Marshall, Fourier transform ion
      cyclotron resonance spectroscopy, *Chem. Phys. Lett. 25*,
      282 (1974).

A37.  M. B. Comisarow and A. G. Marshall, Frequency-sweep
      Fourier transform ion cyclotron resonance spectroscopy,
      *Chem. Phys. Lett. 26*, 489 (1974).

A38.  M. B. Comisarow and A. G. Marshall, Selective-phase ion
      cyclotron resonance spectroscopy, *Can. J. Chem. 52*, 1997
      (1974).

A39.  M. B. Comisarow and A. G. Marshall, Fourier transform ion
      cyclotron resonance spectroscopy, in P. Ausloos (ed.),
      *Interactions Between Ions and Molecules*, Plenum, New York
      (1975), Vol. 6 in N.A.T.O. Advanced Study Institute Series
      B (Physics).

A40.  C. A. Lieder and J. I. Brauman, Detection of neutral
      products in gas-phase, ion-molecule reactions, *J. Amer.
      Chem. Soc. 96*, 4028 (1974).

A41.   J. R. Eyler and G. H. Atkinson, Dye laser-induced photo-
       detachment of electrons from SH⁻ studied by ion cyclotron
       resonance spectroscopy, *Chem. Phys. Lett. 28*, 217 (1974).
A42.   J. R. Eyler, Intracavity dye laser technique for the study
       of laser-induced ionic processes, *Rev. Sci. Instrum. 45*,
       1154 (1974).
A43.   R. T. McIver, Jr., E. B. Ledford, Jr., and J. S. Miller,
       Proposed method for mass spectrometric analysis of ultra-
       low vapor pressure compounds, *Anal. Chem. 47*, 692 (1975).
A44.   T. F. Knott and M. Riggin, Analysis of ion motion in ion
       cyclotron resonance cells, *Can. J. Phys. 52*, 426 (1974).
A45.   M. Bloom and M. Riggin, Theory of ion cyclotron resonance,
       *Can. J. Phys. 52*, 436 (1974).
A46.   M. Riggin and I. B. Woods, Dependence of ion cyclotron
       resonance on electrostatic potentials, *Can. J. Phys. 52*,
       456 (1974).


## ANALYSIS OF ION ENERGIES

B1.    J. L. Beauchamp and S. E. Buttrill, Jr., Proton affinities
       of $H_2S$ and $H_2O$, *J. Chem. Phys. 48*, 1783 (1968).
B2.    J. I. Brauman and L. Blair, Gas phase acidities of amines,
       *J. Amer. Chem. Soc. 91*, 2126 (1969).
B3.    J. I. Brauman and L. K. Blair, Gas-phase acidities of
       alcohols: effects of alkyl groups, *J. Amer. Chem. Soc.
       90*, 6561 (1968).
B4.    D. Holtz and J. L. Beauchamp, Relative basicity of
       phosphine and ammonia in the gas phase, *J. Amer. Chem.
       Soc. 91*, 5913 (1969).
B5.    P. Kriemler and S. E. Buttrill, Jr., Positive and negative
       ion-molecule reactions and proton affinity of ethyl
       nitrite, *J. Amer. Chem. Soc. 92*, 1123 (1970).
B6.    F. Kaplan, P. Cross, and R. Prinstein, Gas-phase stabil-
       ities of bicyclic cations, *J. Amer. Chem. Soc. 92*, 1445
       (1970).
B7.    J. I. Brauman and L. Blair, Gas-phase acidities of
       alcohols, *J. Amer. Chem. Soc. 92*, 5986 (1970).
B8.    D. H. McDaniel, N. Coffman, and J. Strong, Proton affinity
       of trimethylphosphine, *J. Amer. Chem. Soc. 92*, 6697
       (1970).
B9.    H. H. Jaffé and S. Billets, Electrical effect of free
       radical groups, *J. Amer. Chem. Soc. 92*, 6965 (1970).
B10.   D. Holtz, J. L. Beauchamp, and J. R. Eyler, Acidity,
       basicity and ion-molecule reactions of phosphine in the
       gas phase by ion cyclotron resonance spectroscopy, *J.
       Amer. Chem. Soc. 92*, 7045 (1970).

B11.  M. T. Bowers and D. D. Elleman, On the relative proton
      affinity of argon and deuterium, *J. Amer. Chem. Soc. 92*,
      7258 (1970).

B12.  W. B. Nixon and M. M. Bursey, Relative nitryl ion affini-
      ties of some gaseous oxygen Lewis bases by ion cyclotron
      resonance, *Tetra. Lett.*, 4389 (1970).

B13.  J. I. Brauman and L. K. Blair, Gas-phase acidities of
      amines, *J. Amer. Chem. Soc. 93*, 3911 (1971).

B14.  J. I. Brauman, J. M. Riveros, and L. K. Blair, Gas-phase
      basicities of amines, *J. Amer. Chem. Soc. 93*, 3914 (1971).

B15.  M. T. Bowers, D. H. Aue, H. M. Webb, and R. T. McIver,
      Jr., Equilibrium constants for gas-phase ionic reactions.
      Accurate determination of relative proton affinities, *J.
      Amer. Chem. Soc. 93*, 4314 (1971).

B16.  J. I. Brauman and L. K. Blair, Alkyl substituent effects
      on gas-phase acidities. The influence of hydridization,
      *J. Amer. Chem. Soc. 93*, 4315 (1971).

B17.  J. I. Brauman, J. R. Eyler, L. K. Blair, M. J. White,
      M. B. Comisarow, and K. C. Smyth, Gas-phase acidities of
      binary hydrides, *J. Amer. Chem. Soc. 93*, 6360 (1971).

B18.  R. T. McIver, Jr., and J. R. Eyler, Accurate relative
      acidities in the gas phase. Hydrogen sulfide and hydrogen
      cyanide, *J. Amer. Chem. Soc. 93*, 6334 (1971).

B19.  D. Holtz, J. L. Beauchamp, W. G. Henderson, and R. W.
      Taft, Basicity of nitrogen trifluoride in the gas phase by
      ion cyclotron resonance, *Inorg. Chem. 10*, 201 (1971).

B20.  D. Holtz and J. L. Beauchamp, Nitrogen and carbon monoxide
      as nucleophilic reagents in gas phase displacement
      reactions. A novel means of nitrogen fixation, *Nature
      Phys. Science 231*, 204 (1971).

B21.  J. L. Beauchamp and D. Holtz, Xenon as a nucleophile in
      gas-phase displacement reactions: formation of the methyl
      xenonium ion, *Science 173*, 1237 (1971).

B22.  M. Taagepera, W. G. Henderson, R. T. C. Brownlee, J. L.
      Beauchamp, D. Holtz, and R. W. Taft, Gas-phase basicities
      and pyridine substituent effects, *J. Amer. Chem. Soc. 94*,
      1369 (1972).

B23.  E. M. Arnett, F. M. Jones, III, M. Taagepera, W. G.
      Henderson, J. L. Beauchamp, D. Holtz, and R. W. Taft, A
      complete thermodynamic analysis of the anomalous order of
      amine basicities in solution, *J. Amer. Chem. Soc. 94*,
      4724 (1972).

B24.  D. H. Aue, H. M. Webb, and M. T. Bowers, Quantitative
      relative gas-phase basicities of alkylamines. Correlation
      with solution basicity, *J. Amer. Chem. Soc. 94*, 4726
      (1972).

B25. W. G. Henderson, M. Taagepera, D. Holtz, R. T. McIver, Jr., J. L. Beauchamp, and R. W. Taft, Methyl substituent effects in protonated aliphatic amines and their radical cations, *J. Amer. Chem. Soc.* **94**, 4728 (1972).

B26. T. B. McMahon, R. J. Blint, D. P. Ridge, and J. L. Beauchamp, Determination of carbonium ion stabilities by ion cyclotron resonance spectroscopy, *J. Amer. Chem. Soc.* **94**, 8934 (1972).

B27. D. A. Dixon, D. Holtz, and J. L. Beauchamp, Acidity, basicity and gas-phase ion chemistry of hydrogen selenide by ion cyclotron resonance spectroscopy, *Inorg. Chem.* **11**, 960 (1972).

B28. J. M. Riveros and P. W. Tiedemann, Gas phase kinetic protonation site of formamide, *Anais da Academia Brasileira de Ciencias* **44**, 413 (1972).

B29. L. K. Blair, P. C. Isolani, and J. M. Riveros, Formation, reactivity and relative stability of clustered alkoxide ions by ion cyclotron resonance spectroscopy, *J. Amer. Chem. Soc.* **95**, 1057 (1973).

B30. P. Kriemler and S. E. Buttrill, Jr., Ion-molecule reactions and the proton affinities of the nitroalkanes. I. Nitromethane and nitroethane, *J. Amer. Chem. Soc.* **95**, 1365 (1973).

B31. D. H. Aue, H. M. Webb, and M. T. Bowers, Quantitative evaluation of intramolecular strong hydrogen bonding in the gas phase, *J. Amer. Chem. Soc.* **95**, 2699 (1973).

B32. R. T. McIver, Jr., J. A. Scott, and J. M. Riveros, The effect of solvation on the intrinsic relative acidity of methanol and ethanol, *J. Amer. Chem. Soc.* **95**, 2706 (1973).

B33. R. W. Taft, M. Taagepera, K. D. Summerhays, and J. Mitsky, Regarding heats of solution of gaseous anilinium and pyridinium ions in water and intrinsic basicities in aqueous solution, *J. Amer. Chem. Soc.* **95**, 3811 (1973).

B34. J. M. Riveros, A. C. Breda, and L. K. Blair, Formation and relative stability of chloride ion clusters in the gas phase by ion cyclotron resonance spectroscopy, *J. Amer. Chem. Soc.* **95**, 4066 (1973).

B35. J. H. J. Dawson, W. G. Henderson, R. M. O'Malley, and K. R. Jennings, Halide ion transfer reactions in the gas phase ion chemistry of haloalkanes, *Int. J. Mass Spectrom. Ion Phys.* **11**, 61 (1973).

B36. J. M. Riveros, P. W. Tiedemann, and A. C. Breda, Formation of $XeCl^-$ in the gas phase, *Chem. Phys. Lett.* **20**, 345 (1973).

B37. P. C. Isolani, J. M. Riveros, and P. W. Tiedemann, Gas phase proton affinities of carbonyl compounds by ion cyclotron resonance spectroscopy, *J. Chem. Soc., Faraday Trans. II, 69*, 1023 (1973).

B38.  R. T. McIver, Jr., and J. H. Silvers, Gas phase acidity of monosubstituted phenols, *J. Amer. Chem. Soc. 95*, 8462 (1973).

B39.  J. C. Haartz and D. H. McDaniel, Fluoride ion affinity of some Lewis acids, *J. Amer. Chem. Soc. 95*, 8562 (1973).

B40.  J. L. Beauchamp, Reaction mechanisms of organic and inorganic ions in the gas phase, in P. Ausloos (ed.), *Interactions Between Ions and Molecules*, Plenum, New York (1975), Vol. 6 in N.A.T.O. Advanced Study Institute Series B (Physics).

B41.  R. T. McIver, Jr., and J. S. Miller, Thermochemistry of aliphatic alcohols determined by gas-phase ionic equilibria, *J. Amer. Chem. Soc. 96*, 4323 (1974).

B42.  R. H. Staley and J. L. Beauchamp, Basicities and ion-molecule reactions of the methylphosphines in the gas phase by ion cyclotron resonance spectroscopy, *J. Amer. Chem. Soc. 96*, 6252 (1974).

B43.  J. I. Brauman and L. K. Blair, Gas-phase acidities of carbon acids, *J. Amer. Chem. Soc. 90*, 5636 (1968).

B44.  W. J. Hehre, R. T. McIver, Jr., J. A. Pople, and P. v. R. Schleyer, Alkyl substituent effects on the stability of protonated benzene, *J. Amer. Chem. Soc. 96*, 7162 (1974).

B45.  M. T. Bowers and T. Su, Thermal energy ion-molecule reactions, *Adv. Electron. Electron Phys. 34*, 223 (1973).

B46.  D. L. Smith and J. H. Futrell, Internal energy effects on the reaction of $D_3^+$ with $CH_4$, *J. Phys. B. Atomic and Molecular Physics 8*, 803 (1975).

B47.  D. H. Aue, H. M. Webb, and M. T. Bowers, A thermodynamic analysis of solvation effects on the basicities of alkylamines. An electrostatic analysis of substituent effects, *J. Amer. Chem. Soc. 98*, 318 (1976).

B48.  B. S. Freiser and J. L. Beauchamp, Gas phase proton affinities of molecules in excited electronic states by ion cyclotron resonance spectroscopy, *J. Amer. Chem. Soc. 98*, 265 (1976).

B49.  J. F. Wolf, J. L. Devlin, III, R. W. Taft, M. Wolfsberg, and W. J. Hehre, Isotope effects on gas phase reaction processes. I. The determination of equilibrium isotope effects in ion cyclotron resonance spectroscopy, *J. Amer. Chem. Soc. 98*, 287 (1976).

B50.  D. H. Aue, H. M. Webb, and M. T. Bowers, Quantitative proton affinities, ionization potentials, and hydrogen affinities of alkylamines, *J. Amer. Chem. Soc. 98*, 311 (1976).

ANALYSIS OF RATES

C1.   J. L. Beauchamp, Theory of collision-broadened ion
      cyclotron resonance spectra, *J. Chem. Phys. 46*, 1231
      (1967).
C2.   R. C. Dunbar, Energy dependence of ion-molecule reactions,
      *J. Chem. Phys. 47*, 5445 (1967).
C3.   J. King, Jr., and D. D. Elleman, Charge exchange reactions
      in xenon-methane mixtures, *J. Chem. Phys. 48*, 4803 (1968).
C4.   M. T. Bowers, D. D. Elleman, and J. King, Jr., Kinetic
      analysis of the ion-molecule reactions in nitrogen-
      hydrogen mixtures using ion cyclotron resonance, *J. Chem.
      Phys. 50*, 1840 (1969).
C5.   S. E. Buttrill, Jr., Measurement of ion-molecule reaction
      rate constants using ion cyclotron resonance, *J. Chem.
      Phys. 50*, 4125 (1969).
C6.   M. T. Bowers, D. D. Elleman, and J. King, Jr., Analysis of
      the ion-molecule reactions in gaseous $H_2$, $D_2$ and HD by ion
      cyclotron resonance techniques, *J. Chem. Phys. 50*, 4787
      (1969).
C7.   R. P. Clow and J. H. Futrell, Observation of charge
      exchange in xenon-methane mixtures by ion cyclotron double
      resonance, *J. Chem. Phys. 50*, 5041 (1969).
C8.   J. Schaefer and J. M. S. Henis, Dependence of ion-molecule
      reaction cross section on internal energy, *J. Chem. Phys.
      51*, 4671 (1969).
C9.   M. T. Bowers and D. D. Elleman, Kinetic analysis of the
      concurrent ion-molecule reactions in mixtures of argon and
      nitrogen with $H_2$, $D_2$ and HD utilizing ion-ejection ion
      cyclotron resonance techniques, *J. Chem. Phys. 51*, 4606
      (1969).
C10.  M. Inoue and S. Wexler, Isotopic exchange in $CH_4$-$D_2$ and
      $CD_4$-$H_2$ mixture studied by ion cyclotron resonance
      spectroscopy.  The mechanism of self-induced labeling of
      methane by tritium, *J. Amer. Chem. Soc. 91*, 5730 (1969).
C11.  W. T. Huntress and J. L. Beauchamp, Application of ion
      cyclotron resonance to the study of ionizing reactions of
      metastable molecules, *Int. J. Mass Spectrom. Ion Phys. 3*,
      149 (1969).
C12.  L. R. Anders, Study of the energetics of ion-molecule
      reactions by pulsed ion cyclotron double resonance, *J.
      Phys. Chem. 73*, 469 (1969).
C13.  R. C. Dunbar, Energy dependence of methanol proton trans-
      fer reaction rate, *J. Chem. Phys. 52*, 2780 (1970).
C14.  S. E. Buttrill, Jr., Calculation of ion-molecule reaction
      product distributions using the quasiequilibrium theory of
      mass spectra, *J. Chem. Phys. 52*, 6174 (1970).

C15.    A. G. Marshall and S. E. Buttrill, Jr., Calculation of
        ion-molecule reaction rate constants from ion cyclotron
        resonance spectra:  methyl fluoride, *J. Chem. Phys.* *52*,
        2752 (1970).

C16.    M. Mosesman and W. T. Huntress, On the reaction of $O^+$ and
        $CO_2$, *J. Chem. Phys.* *53*, 462 (1970).

C17.    J. M. S. Henis and C. A. Mabie, Determination of auto-
        ionization lifetimes by ion cyclotron resonance line-
        widths, *J. Chem. Phys.* *53*, 2999 (1970).

C18.    M. T. Bowers and D. D. Elleman, Analysis of the ion-
        molecule reactions in hydrogen-methane mixtures using ion
        cyclotron resonance, *J. Amer. Chem. Soc.* *92*, 1847 (1970).

C19.    A. A. Herod, A. G. Harrison, R. M. O'Malley, A. J. Ferrer-
        Correia, and K. R. Jennings, A comparison of the zero-
        field pulsing technique and the ICR technique for studying
        ion-molecule reactions, *J. Phys. Chem.* *74*, 2720 (1970).

C20.    R. P. Clow and J. H. Futrell, Ion-cyclotron resonance
        study of the kinetic energy dependence of ion-molecule re-
        action rates.  I.  Methane, hydrogen and rare gas-hydrogen
        systems, *Int. J. Mass Spectrom. Ion Phys.* *4*, 165 (1970).

C21.    G. C. Goode, A. J. Ferrer-Correia, and K. R. Jennings, The
        interpretation of double resonance signals in ion cyclo-
        tron resonance mass spectrometry, *Int. J. Mass Spectrom.
        Ion Phys.* *5*, 229 (1970).

C22.    G. C. Goode, R. M. O'Malley, A. J. Ferrer-Correia, R. I.
        Massey, K. R. Jennings, J. H. Futrell, and P. M.
        Llewellyn, Rate constants for ion-molecule reactions
        determined by ICR mass spectrometry, *Int. J. Mass
        Spectrom. Ion Phys.* *5*, 393 (1970).

C23.    W. T. Huntress, M. M. Mosesman, and D. D. Elleman,
        Relative rates and their dependence on kinetic energy for
        ion-molecule reactions in ammonia, *J. Chem. Phys.* *54*, 843
        (1971).

C24.    M. L. Gross and J. Norbeck, Effects of vibrational energy
        on the rates of ion-molecule reactions, *J. Chem. Phys.* *54*,
        3651 (1971).

C25.    M. B. Comisarow, Comprehensive theory for ion cyclotron
        resonance power absorption:  application to lineshapes for
        reactive and non-reactive ions, *J. Chem. Phys.* *55*, 205
        (1971).

C26.    A. G. Marshall, Theory for ion cyclotron resonance
        absorption line shapes, *J. Chem. Phys.* *55*, 1343 (1971).

C27.    W. T. Huntress, Ion cyclotron resonance power absorption:
        collision frequencies for $CO_2^+$, $N_2^+$ and $H_3^+$ ions in their
        parent gases, *J. Chem. Phys.* *55*, 2146 (1971).

C28.    M. T. Bowers and J. B. Laudenslager, Mechanism of charge
        transfer reactions:  reactions of rare gas ions with the

*trans-*, *cis-*, and 1,1-difluoroethylene geometric isomers, *J. Chem. Phys. 56*, 4711 (1972).

C29.   C. A. Lieder, R. W. Wien, and R. T. McIver, Jr., Ion-molecule collision frequencies in gases determined by phase coherent pulsed ICR, *J. Chem. Phys. 56*, 5184 (1972).

C30.   P. P. Dymerski and R. C. Dunbar, ICR study of non-reactive collision rate constants, *J. Chem. Phys. 57*, 4049 (1972).

C31.   R. P. Clow and J. H. Futrell, Ion-molecule reactions in isotopic hydrogen by ion cyclotron resonance, *Int. J. Mass Spectrom. Ion Phys. 8*, 119 (1972).

C32.   T. McAllister, Comparison of expressions for the determination of rate constants of ion-molecule reactions by ion cyclotron resonance mass spectrometry, *Int. J. Mass Spectrom. Ion Phys. 8*, 162 (1972).

C33.   T. McAllister, Ion-molecule reaction kinetics by ion cyclotron resonance mass spectrometry, *Int. J. Mass Spectrom. Ion Phys. 9*, 127 (1972).

C34.   T. E. Sharp, J. R. Eyler, and E. Li, Trapped-ion motion in ion cyclotron resonance spectroscopy, *Int. J. Mass Spectrom. Ion Phys. 9*, 421 (1972).

C35.   R. Marx and G. Mauclaire, Ion cyclotron resonance study of positive and negative ion-molecule reactions in ammonia, *Int. J. Mass Spectrom. Ion Phys. 10*, 213 (1972).

C36.   D. L. Smith and J. H. Futrell, Ion-molecule reactions in the $CO_2/H_2$ system by ion cyclotron resonance, *Int. J. Mass Spectrom. Ion Phys. 10*, 405 (1972).

C37.   K. A. G. MacNeil and J. Futrell, Ion-molecule reactions in gaseous acetone, *J. Phys. Chem. 76*, 409 (1972).

C38.   L. Kevan and J. H. Futrell, Determination of collision rate constants of fluorocarbon ions by ion cyclotron resonance, *J. Chem. Soc., Faraday Trans. II, 68*, 1742 (1972).

C39.   M. T. Bowers and D. D. Elleman, Thermal energy charge transfer reactions of rare-gas ions to methane, ethane, propane and silane. The importance of Franck-Condon factors, *Chem. Phys. Lett. 16*, 486 (1972).

C40.   M. L. Gross, Ion cyclotron resonance spectrometry. A means of evaluating kinetic shift, *Org. Mass Spectrom. 6*, 827 (1972).

C41.   T. Su and M. T. Bowers, Theory of ion-polar molecule collisions: comparison with experimental charge transfer reactions of rare gas ions to geometric isomers of difluorobenzene and dichloroethylene, *J. Chem. Phys. 58*, 3027 (1973).

C42.   M. T. Bowers, T. Su, and V. G. Anicich, Theory of ion-polar molecule collisions. Kinetic energy dependence of ion-polar molecule reactions: $CH_3OH^{+\cdot} + CH_3OH \rightarrow CH_3OH_2^+ +$ $\cdot CH_3O$, *J. Chem. Phys.* *58*, 5175 (1973).

C43.  D. L. Smith and J. H. Futrell, Low energy study of symmetric and asymmetric charge-transfer reactions, *J. Chem. Phys.* *59*, 463 (1973).

C44.  J. H. Futrell, Ion molecule reactions in perdeuterio-methane, *J. Chem. Phys.* *59*, 4061 (1973).

C45.  T. McAllister, Ion-molecule reactions in mixtures of $N_2O$ with $H_2$ and $CH_4$, *Int. J. Mass Spectrom. Ion Phys.* *10*, 419 (1973).

C46.  V. G. Anicich and M. T. Bowers, Absolute ion-molecule rate constants from drift cell ion cyclotron resonance spectroscopy, *Int. J. Mass Spectrom. Ion Phys.* *11*, 329 (1973).

C47.  W. T. Huntress, On ion molecule reactions in ammonia, *Int. J. Mass Spectrom. Ion Phys.* *11*, 495 (1973).

C48.  W. T. Huntress and M. T. Bowers, Reactions of excited and ground state $H_3^+$ ions with methyl substituted hydrides, *Int. J. Mass Spectrom. Ion Phys.* *12*, 1 (1973).

C49.  B. S. Freiser, T. B. McMahon, and J. L. Beauchamp, Observation of ion ejection phenomena in ion cyclotron double resonance experiments, *Int. J. Mass Spectrom. Ion Phys.* *12*, 249 (1973).

C50.  J. I. Brauman, C. A. Lieder, and M. J. White, Homogeneous catalysis of a gas-phase, ion-molecule reactions, *J. Amer. Chem. Soc.* *95*, 927 (1973).

C51.  T. Su and M. T. Bowers, Ion-polar molecule collisions: proton transfer reactions of $H_3^+$ and $CH_5^+$ to the geometric isomers of difluoroethylene, dichloroethylene and difluorobenzene, *J. Amer. Chem. Soc.* *95*, 1370 (1973).

C52.  T. Su and M. T. Bowers, Ion-polar molecule collisions. Proton transfer reactions of $C_4H_9^+$ ions with $NH_3$, $CH_3NH_2$, $(C_2H_5)_2NH$, and $(CH_3)_3N$, *J. Amer. Chem. Soc.* *95*, 7611 (1973).

C53.  T. Su and M. T. Bowers, Ion-polar molecule collisions. The effect of molecular size on ion-polar molecule rate constants, *J. Amer. Chem. Soc.* *95*, 7609 (1973).

C54.  V. G. Anicich and M. T. Bowers, Reactions of the parent ions in vinyl chloride and vinyl fluoride as studied by ion cyclotron resonance, *Int. J. Mass Spectrom. Ion Phys.* *12*, 231 (1973).

C55.  T. Su and M. T. Bowers, Ion-polar molecule collisions: the effect of ion size on ion-polar molecule rate constants; the parameterization of the average-dipole-orientation theory, *Int. J. Mass Spectrom. Ion Phys.* *12*, 347 (1973).

C56.  M. T. Bowers, W. J. Chesnavich, and W. T. Huntress, Jr., Deactivation of internally excited $H_3^+$ ions: comparison of experimental product distribution of reactions of $H_3^+$ ions with $CH_3NH_2$, $CH_3OH$, and $CH_3SH$ with predictions of quasiequilibrium theory calculations, *Int. J. Mass Spectrom. Ion Phys.* *12*, 357 (1973).

C57.  S. E. Buttrill, Jr., Temperature dependence of the rates of ion-molecule collisions, *J. Chem. Phys.* *58*, 656 (1973).

C58.  Sigmund Jaffé, Ze'ev Karplus, and Fritz S. Klein, Ion cyclotron mass spectrometric study of the reaction $N_2^+ + N_2 \rightarrow N_3^+ + N$, *J. Chem. Phys.* *58*, 2190 (1973).

C59.  G. W. Stewart, J. M. S. Henis, and P. P. Gaspar, Implications of ion-molecule reactions observed in silane for recoil silicon atom studies, *J. Chem. Phys.* *58*, 890 (1973).

C60.  J. M. S. Henis, G. W. Stewart, and P. P. Gaspar, Endothermic ion-molecule reactions in silane, *J. Chem. Phys.* *58*, 3639 (1973).

C61.  W. T. Huntress, Jr., and R. F. Pinizzotto, Jr., Product distributions and rate constants for ion-molecule reactions in water, hydrogen sulfide, ammonia, and methane, *J. Chem. Phys.* *59*, 4742 (1973).

C62.  P. R. Kemper and M. T. Bowers, Mechanism of thermal energy gas phase charge exchange reaction: $He^{+\cdot} + N_2$, *J. Chem. Phys.* *59*, 4915 (1973).

C63.  W. J. van der Hart, Calculation of rate constants from ion cyclotron resonance spectra, *Chem. Phys. Lett.* *23*, 93 (1973).

C64.  P. P. Dymerski, R. C. Dunbar, and J. V. Dugan, Jr., ICR study of nonreactive ion-molecule collision rate constants for $Cl^-$ and $Cr(CO)_5^-$, *J. Chem. Phys.* *61*, 298 (1974).

C65.  M. D. Sefcik, J. M. S. Henis, and P. P. Gaspar, Silanium $SiH_5^+$. The protonation of silane and the chemistry of silanium ions studied by ion cyclotron resonance spectroscopy, *J. Chem. Phys.* *61*, 4329 (1974).

C66.  J. B. Laudenslager, W. T. Huntress, Jr., and M. T. Bowers, Near thermal energy charge transfer reactions of rare gas ions with diatomic and simple polyatomic molecules: the importance of Franck-Condon factors and energy resonance on the magnitude of the rate constants, *J. Chem. Phys.* *61*, 4600 (1974).

C67.  J. M. S. Henis, M. K. Tripodi, and M. D. Sefcik, Effect of charge redistribution on ion-molecule reaction rates, *J. Amer. Chem. Soc.* *96*, 1660 (1974).

C68.  R. W. Odom, D. L. Smith, and J. H. Futrell, A new measurement of the $SF_6^-$ auto-ionization lifetime, *Chem. Phys. Lett.* *24*, 227 (1974).

C69.  M. S. Foster and J. L. Beauchamp, Electron attachment to
      sulfur hexafluoride:  formation of stable $SF_6^-$ at low
      pressure, *Chem. Phys. Lett.* *31*, 482 (1975).

C70.  M. Riggin, Ion cyclotron resonance collision frequencies
      in alkali ions in rare gases, *Can. J. Phys.* *52*, 1683
      (1974).


STRUCTURES OF IONS AND NEUTRALS

D1.   J. Diekman, J. MacLeod, C. Djerassi, and J. D.
      Baldeschwieler, Determination of the structures of the
      ions produced in the single and double McLafferty
      rearrangements by ion cyclotron resonance spectroscopy, *J.
      Amer. Chem. Soc.* *91*, 2069 (1969).

D2.   G. Eadon, J. Diekman, and C. Djerassi, Application of ion
      cyclotron resonance to the structure elucidation of the
      $C_3H_6O^{+\cdot}$ ion formed in the double McLafferty rearrangement,
      *J. Amer. Chem. Soc.* *91*, 3986 (1969).

D3.   G. Eadon, J. Diekman, and C. Djerassi, Application of ion
      cyclotron resonance to the structure elucidation of the
      $C_3H_6O^{+\cdot}$ ion formed in the double McLafferty rearrangement,
      *J. Amer. Chem. Soc.* *92*, 6205 (1970).

D4.   J. L. Beauchamp and R. C. Dunbar, Identification of $C_2H_5O^+$
      structural isomers by ion-cyclotron resonance spectros-
      copy, *J. Amer. Chem. Soc.* *92*, 1477 (1970).

D5.   M. L. Gross and F. W. McLafferty, Identification of $C_3H_6^{+\cdot}$
      structural isomers by ion cyclotron resonance spectros-
      copy, *J. Amer. Chem. Soc.* *93*, 1267 (1971).

D6.   M. T. Bowers and P. R. Kemper, Analysis of the mechanism
      of reaction of $H_3^+$ with ethylene oxide and acetaldehyde,
      *J. Amer. Chem. Soc.* *93*, 5352 (1971).

D7.   M. K. Hoffman and M. M. Bursey, The structure of the
      molecular ion of $C_7H_8$ isomers:  an ICR study, *Tetra.
      Lett.*, 2539 (1971).

D8.   M. L. Gross, P. H. Lin, and S. J. Franklin, Analytical
      applications of ion-molecule reactions.  Identification of
      $C_5H_{10}$ isomers by ion cyclotron resonance spectrometry,
      *Anal. Chem.* *44*, 974 (1972).

D9.   H. H. Jaffé and S. Billets, On the structure of protonated
      ethylene, *J. Amer. Chem. Soc.* *94*, 674 (1972).

D10.  S. A. Benezra and M. M. Bursey, Steric inhibition of
      gaseous ionic acetylation, *J. Amer. Chem. Soc.* *94*, 1024
      (1972).

D11.  D. J. McAdoo, F. W. McLafferty, and P. F. Bente, III,
      Ion cyclotron resonance spectroscopy in structure

determination. II. Propyl ions, *J. Amer. Chem. Soc. 94*, 2027 (1972).

D12. M. L. Gross, An ion cyclotron resonance study of the structure of $C_3H_6^+$ and the mechanism of its reaction with ammonia, *J. Amer. Chem. Soc. 94*, 3744 (1972).

D13. K. B. Tomer and C. Djerassi, Mass spectrometry in structural and stereochemical problems CCXXVIII. Application of ion cyclotron resonance for differentiation between tautomers: vinylthiol and thioacetaldehyde, *J. Amer. Chem. Soc. 95*, 5335 (1973).

D14. C. L. Wilkins and M. L. Gross, A study of the styrene ion-molecule reaction by ion cyclotron resonance, *J. Amer. Chem. Soc. 93*, 895 (1971).

D15. P. K. Pearson, H. F. Shaefer, III, J. H. Richardson, L. M. Stephenson, and J. I. Brauman, Three isomers of the $NO_2^-$ ion, *J. Amer. Chem. Soc. 96*, 6778 (1974).

D16. R. T. McIver, Jr., Use of ion-molecule reaction equilibria to identify structural isomers, *Org. Mass Spectrom. 10*, 396 (1975).


FUNCTIONAL GROUP CHEMISTRY

E1. J. L. Beauchamp, L. R. Anders, and J. D. Baldeschwieler, The study of ion-molecule reactions in chloroethylene by ion cyclotron resonance spectroscopy, *J. Amer. Chem. Soc. 89*, 4569 (1967).

E2. J. M. S. Henis, An ion cyclotron resonance study of ion molecule reactions in methanol, *J. Amer. Chem. Soc. 90*, 844 (1968).

E3. G. A. Gray, Study of ion-molecule reaction mechanisms in acetonitrile by ion cyclotron resonance, *J. Amer. Chem. Soc. 90*, 2177 (1968).

E4. F. Kaplan, Identification of collision-induced fragmentation pathways by ion cyclotron double resonance, *J. Amer. Chem. Soc. 90*, 4483 (1968).

E5. R. C. Dunbar, Ion-molecule chemistry of diborane by ion cyclotron resonance, *J. Amer. Chem. Soc. 90*, 5676 (1968).

E6. G. A. Gray, Study of ion-molecule reactions and reaction mechanisms in acetonitrile by ion cyclotron resonance, *J. Amer. Chem. Soc. 90*, 6002 (1968).

E7. J. King and D. Elleman, Ion cyclotron resonance study of the secondary ion-molecule reactions in hexafluoroethane, *J. Chem. Phys. 48*, 412 (1968).

E8. J. Schaefer and J. M. S. Henis, Electron density rearrangement description of ion-molecule reactions, *J. Chem. Phys. 49*, 5377 (1968).

E9.   M. T. Bowers, D. D. Elleman, and J. L. Beauchamp, Ion
      cyclotron resonance of olefins I.  A study of the ion-
      molecule reactions in electron impacted ethylene, *J. Phys.
      Chem. 72*, 3599 (1968).

E10.  J. L. Beauchamp, Ionic dehydration of aliphatic alcohols
      in the gas phase, *J. Amer. Chem. Soc. 91*, 5925 (1969).

E11.  J. M. S. Henis, Analytical implications of ion cyclotron
      resonance spectroscopy, *Anal. Chem. 41* (10), 22A (1969).

E12.  W. T. Huntress, J. D. Baldeschwieler, and C. Ponnamperuma,
      Ion-molecule reactions in hydrogen cyanide, *Nature 223*,
      468 (1969).

E13.  R. M. O'Malley and K. R. Jennings, Ion cyclotron resonance
      mass spectrometry of acetylene, *Int. J. Mass Spectrom. Ion
      Phys. 2*, 257 (1969).

E14.  R. M. O'Malley and K. R. Jennings, Ion cyclotron resonance
      mass spectra of fluoroalkenes I.  Ion-molecule reactions
      of ethylene and vinyl fluoride, *Int. J. Mass Spectrom. Ion
      Phys. 2*, 441 (1969).

E15.  S. E. Buttrill, Jr., Ion-molecule reactions of hydrogen
      sulfide with ethylene and acetylene, *J. Amer. Chem. Soc.
      92*, 3560 (1970).

E16.  W. T. Huntress and D. D. Elleman, An ion cyclotron reso-
      nance study of ion-molecule reactions in methane-ammonia
      mixtures, *J. Amer. Chem. Soc. 92*, 3565 (1970).

E17.  S. A. Benezra, M. K. Hoffman, and M. M. Bursey, Electro-
      philic aromatic substitution reactions.  An ion cyclotron
      resonance study, *J. Amer. Chem. Soc. 92*, 7501 (1970).

E18.  D. Holtz, J. L. Beauchamp, and S. S. Woodgate, Nucleo-
      philic displacement reactions in the gas phase, *J. Amer.
      Chem. Soc. 92*, 7484 (1970).

E19.  S. Billets, H. H. Jaffé, and F. Kaplan, Rearrangements of
      molecular ions of dialkyl-N-nitrosamines, *J. Amer. Chem.
      Soc. 92*, 6964 (1970).

E20.  M. M. Bursey, T. A. Elwood, M. K. Hoffman, T. A. Lehman,
      and J. M. Tesarek, Analytical ion cyclotron resonance
      spectrometry.  Acetylation as a chemical ionization
      technique, *Anal. Chem. 42*, 1370 (1970).

E21.  T. A. Lehman, T. A. Elwood, M. K. Hoffman, and M. M.
      Bursey, Dehydration of the protonated butane-2,3-dione-
      ethanol ion $\left((CH_3COCOCH_3)(C_2H_5OH)H^+\right)$ in the gaseous phase,
      *J. Chem. Soc. (B)*, 1717 (1970).

E22.  J. M. S. Henis, Ion-molecule reactions in olefins, *J.
      Chem. Phys. 52*, 282 (1970).

E23.  J. M. S. Henis, Isotopic exchange in olefin ion-molecule
      reactions, *J. Chem. Phys. 52*, 292 (1970).

E24.  M. T. Bowers, D. D. Elleman, R. M. O'Malley, and K. R.
      Jennings, Analysis of ion-molecule reactions in allene and

propyne by ion cyclotron resonance, *J. Phys. Chem. 74*, 2583 (1970).

E25. M. K. Hoffman, T. A. Elwood, T. A. Lehman, and M. M. Bursey, Mechanism of acetyl transfer to oxygen bases in ion-molecule reactions of 2,3 butandione, *Tetra. Lett.*, 4021 (1970).

E26. J. R. Eyler, Study of ion-molecule reactions in phosphine by ion cyclotron resonance, *Inorg. Chem. 9*, 981 (1970).

E27. T. A. Lehman, T. A. Elwood, J. T. Bursey, M. M. Bursey, and J. L. Beauchamp, Dehydration reactions involving major rearrangements: comparison of ordinary mass spectral fragmentation to ion-molecule reaction processes, *J. Amer. Chem. Soc. 93*, 2108 (1971).

E28. R. C. Dunbar, Positive and negative ions and ion-molecule reactions of several boron hydrides studied by ion cyclotron resonance, *J. Amer. Chem. Soc. 93*, 4167 (1971).

E29. R. C. Dunbar, Photodissociation of the $CH_3Cl^+$ and $N_2O^+$ cations, *J. Amer. Chem. Soc. 93*, 4354 (1971).

E30. M. S. Foster and J. L. Beauchamp, Potential of ion cyclotron resonance spectroscopy for the study of intrinsic properties and reactivity of transition metal complexes in the gas phase. Ion-molecule reactions of iron pentacarbonyl, *J. Amer. Chem. Soc. 93*, 4924 (1971).

E31. D. P. Ridge and J. L. Beauchamp, Chemical consequences of strong hydrogen bonding in the reactions of organic ions in the gas phase. Induced fragmentation of aliphatic alcohols, *J. Amer. Chem. Soc. 93*, 5925 (1971).

E32. M. M. Bursey, M. K. Hoffman, and S. A. Benezra, Ion cyclotron resonance study of the mechanism of the loss of $C_3H_6$ from the molecular ion of n-butylbenzene, *Chem. Comm.*, 1417 (1971).

E33. M. M. Bursey and M. K. Hoffman, An approach to stereochemical analysis by ion cyclotron resonance distinguishing *exo-* and *endo-*norborneol, *Can. J. Chem. 49*, 3395 (1971).

E34. M. Irie, K. Aoyagi, K. Hayashi, J. Sohma, and T. Miyamae, Initiation mechanism of radiation-induced ionic polymerization of isobutene studied by ion cyclotron resonance, *Bull. Chem. Soc. Jap. 44*, 2261 (1971).

E35. M. S. Foster and J. L. Beauchamp, Gas-phase ion chemistry of azomethane by ion cyclotron resonance spectroscopy, *J. Amer. Chem. Soc. 94*, 2425 (1972).

E36. J. L. Beauchamp and M. C. Caserio, Ion-molecule reactions of 2-butanol by ion cyclotron resonance spectroscopy, *J. Amer. Chem. Soc. 94*, 2638 (1972).

E37. J. L. Beauchamp, D. Holtz, S. D. Woodgate, and S. L. Patt, Thermochemical properties and ion-molecule reactions of the alkyl halides in the gas phase by ion cyclotron resonance spectroscopy, *J. Amer. Chem. Soc. 94*, 2798 (1972).

E38.  T. H. Morton and J. L. Beauchamp, Chemical consequences of strong hydrogen bonding in the reactions of organic ions in the gas phase, interaction of remote functional groups, *J. Amer. Chem. Soc. 94*, 3671 (1972).

E39.  R. P. Clow and J. H. Futrell, Ion cyclotron resonance study of the mechanism of chemical ionization. Mass spectroscopy of selected hydrocarbons using methane reagent gas, *J. Amer. Chem. Soc. 94*, 3748 (1972).

E40.  M. T. Bowers, D. H. Aue, and D. D. Elleman, Mechanisms of ion-molecule reactions of propene and cyclopropane, *J. Amer. Chem. Soc. 94*, 4255 (1972).

E41.  J. M. Kramer and R. C. Dunbar, A gas-phase photon-induced ion-molecule reaction studied by ion cyclotron resonance spectroscopy, *J. Amer. Chem. Soc. 94*, 4346 (1972).

E42.  J. R. Hass, M. M. Bursey, D. G. Kingston, and H. P. Tannenbaum, Reketonization of a McLafferty product ion studied by ion-molecule reactivity, *J. Amer. Chem. Soc. 94*, 5095 (1972).

E43.  R. C. Dunbar, J. Shen, and G. A. Olah, Substituent effects in gas-phase ionic nitration and acetylation of aromatics, *J. Amer. Chem. Soc. 94*, 6862 (1972).

E44.  R. C. Dunbar, J. Shen, and G. A. Olah, Ion-molecule reactions of ethane at low electron energy, *J. Chem. Phys. 56*, 3794 (1972).

E45.  W. T. Huntress, Hydrogen atom scrambling in ion-molecule reactions of methane and ethylene, *J. Chem. Phys. 56*, 5111 (1972).

E46.  T. McAllister, Ion cyclotron double resonance of ion-molecule reactions in ethane, *J. Chem. Phys. 56*, 5192 (1972).

E47.  J. M. S. Henis, G. W. Stewart, M. K. Tripodi, and P. P. Gaspar, Ion-molecule reactions in silane, *J. Chem. Phys. 57*, 389 (1972).

E48.  R. C. Dunbar, Ion-molecule reactions of diborane and oxygen-containing compounds, *J. Phys. Chem. 76*, 2467 (1972).

E49.  M. L. Gross, P. Lin, and S. Franklin, Analytical applications of ion-molecule reactions. Identification of $C_5H_{10}$ isomers by ion cyclotron resonance spectrometry, *Anal. Chem. 44*, 974 (1972).

E50.  R. J. Liedtke, A. F. Gerrard, J. Diekman, and C. Djerassi, Mass spectrometry in structural and stereochemical problems CCXV. Behavior of phenyl-substituted $\alpha,\beta$-unsaturated ketones upon electron impact. Promotion of

hydrogen rearrangement processes, *J. Org. Chem. 37*, 776
(1972).

E51.  R. C. Dunbar, Photodissociation of toluene parent cations,
*J. Amer. Chem. Soc. 95*, 472 (1973).

E52.  J. K. Kim, M. C. Findlay, W. G. Henderson, and M. C.
Caserio, Ion cyclotron resonance spectroscopy. Neighboring group effects in the gas-phase ionization of
$\beta$-substituted alcohols, *J. Amer. Chem. Soc. 95*, 2184
(1973).

E53.  R. C. Dunbar and E. Fu, Photodissociation spectroscopy of
gaseous $C_7H_8^+$ cations, *J. Amer. Chem. Soc. 95*, 2716 (1973).

E54.  P. W. Tiedemann and J. M. Riveros, Mechanism of ionic
self-acylation in the gas phase by ion cyclotron resonance
spectroscopy, *J. Amer. Chem. Soc. 95*, 3140 (1973).

E55.  W. T. Huntress, Jr., R. F. Pinizzotto, and J. B.
Laudenslager, Ion-molecule reactions in mixtures of
methane with water, hydrogen sulfide, and ammonia, *J.
Amer. Chem. Soc. 95*, 4107 (1973).

E56.  R. C. Dunbar, Gas-phase photodissociation of alkylbenzene
cations, *J. Amer. Chem. Soc. 95*, 6191 (1973).

E57.  R. C. Dunbar and J. M. Kramer, ICR study of the angular
dependence of photodissociation and the photodissociation
transition moment orientation in gaseous $CH_3Cl^+$, *J. Chem.
Phys. 58*, 1266 (1973).

E58.  R. M. O'Malley, K. R. Jennings, M. T. Bowers, and V. G.
Anicich, ICR mass spectra of fluoroalkenes. II. Ion
molecule reactions in the 1,1-difluoroethylene system,
*Int. J. Mass Spectrom. Ion Phys. 11*, 89 (1973).

E59.  V. G. Anicich, M. T. Bowers, R. M. O'Malley, and K. R.
Jennings, ICR mass spectra of fluoroalkenes. III.
Thermal ion molecule reactions in $C_2HF_3$ and $C_2F_4$ by ion
cyclotron resonance, *Int. J. Mass Spectrom. Ion Phys. 11*,
99 (1973).

E60.  A. J. Ferrer-Correia and K. R. Jennings, ICR mass spectra
of fluoroalkenes. IV. Ion molecule reactions in mixtures
of ethylene and the fluoroethylenes, *Int. J. Mass
Spectrom. Ion Phys. 11*, 111 (1973).

E61.  N. M. M. Nibbering, An ICR study on the structure of the
$C_6H_6O^{+\cdot}$ ion from phenetole, *Tetra. 29*, 385 (1973).

E62.  J. M. Kramer and R. C. Dunbar, Photodissociation of
gaseous olefinic cations, *J. Chem. Phys. 59*, 3092 (1973).

E63.  R. T. McIver, Jr., and J. H. Silvers, Gas phase acidity of
monosubstituted phenols, *J. Amer. Chem. Soc. 95*, 8462
(1973).

E64.  J. D. Henion, M. C. Sammons, C. E. Parker, and M. M.
Bursey, Failure of fundamental rate constant expressions

for the gaseous ionic acetylation of 2-alkylpyridines, *Tetra. Lett.*, 4925 (1973).

E65.  R. C. Dunbar, J. Shen, E. Melby, and G. A. Olah, Gas phase methylenation of benzene and substituted benzenes by $C_2H_5O^+$ ion, a novel electrophilic aromatic substitution, *J. Amer. Chem. Soc. 95*, 7200 (1973).

E66.  P. W. Tiedemann and J. M. Riveros, Ion-molecule reactions of acids and esters with alcohols. Gas-phase analogs of acidic esterification processes, *J. Amer. Chem. Soc. 96*, 185 (1974).

E67.  J. I. Brauman, C. A. Lieder, and W. N. Olmstead, Gas-phase nucleophilic displacement reactions, *J. Amer. Chem. Soc. 96*, 4030 (1974).

E68.  D. P. Ridge and J. L. Beauchamp, Chemical consequences of strong hydrogen bonding in the reactions of organic ions in the gas phase. Base induced elimination reactions, *J. Amer. Chem. Soc. 96*, 637 (1974).

E69.  D. P. Ridge and J. L. Beauchamp, Reactions of strong bases with alkyl halides in the gas phase. A new look at E2 base-induced elimination reactions without solvent participation, *J. Amer. Chem. Soc. 96*, 3595 (1974).

E70.  C. V. Pesheck and S. E. Buttrill, Jr., Ion cyclotron resonance studies of the chemical ionization of esters, *J. Amer. Chem. Soc. 96*, 6027 (1974).

E71.  M. S. Foster and J. L. Beauchamp, Gas-phase ion chemistry of iron pentacarbonyl by ion cyclotron resonance spectroscopy. New insights into the properties and reactions of transition metal complexes in the absence of complicating solvation phenomena, *J. Amer. Chem. Soc. 97*, 4808 (1975).

E72.  M. S. Foster and J. L. Beauchamp, Ion-molecule reactions and gas-phase basicity of ferrocene, *J. Amer. Chem. Soc. 97*, 4814 (1975).

E73.  E. W. Fu, P. P. Dymerski, and R. C. Dunbar, The photo-dissociation and high resolution laser photodissociation of halogen-substituted toluene cations, *J. Amer. Chem. Soc. 98*, 337 (1976).


NEGATIVE IONS

F1.  J. I. Brauman and K. C. Smyth, Photodetachment energies of negative ions by ion cyclotron resonance spectroscopy. Electron affinities of neutral radicals, *J. Amer. Chem. Soc. 91*, 7778 (1969).

F2.  K. C. Smyth, R. T. McIver, Jr., J. I. Brauman, and R. W. Wallace, Photodetachment of negative ions using a continuously tunable laser and an ion cyclotron resonance spectrometer, *J. Chem. Phys. 54*, 2758 (1971).

F3.   K. C. Smyth and J. I. Brauman, Photodetachment of electrons from phosphide ion; the electron affinity of $PH_2$, *J. Chem. Phys. 56*, 1132 (1972).

F4.   K. C. Smyth and J. I. Brauman, Photodetachment of electrons from amide and arsenide ions: the electron affinities of $NH_2 \cdot$ and $AsH_2 \cdot$, *J. Chem. Phys. 56*, 4620 (1972).

F5.   K. C. Smyth and J. I. Brauman, Photodetachment of an electron from selenide ion; the electron affinity and spin-orbit coupling constant for SeH, *J. Chem. Phys. 56*, 5993 (1972).

F6.   T. McAllister, Negative ion production by secondary electrons in a mass spectrometer ion source, *Chem. Commu.*, 245 (1972).

F7.   J. C. Haartz and D. H. McDaniel, Fluoride ion affinity of some Lewis acids, *J. Amer. Chem. Soc. 95*, 8562 (1973).

F8.   R. Marx, G. Mauclaire, F. C. Fehsenfeld, D. B. Dunkin, and E. E. Ferguson, Negative ion reactions in $N_2O$ at low energies, *J. Chem. Phys. 58*, 3267 (1973).

F9.   J. H. Richardson, L. M. Stephenson, and J. I. Brauman, Photodetachment of electrons from large molecular systems: cyclopentadienide and methylcyclopentadienide ions. An upper limit to the electron affinities of $C_5H_5 \cdot$ and $CH_3C_5H_4 \cdot$, *J. Chem. Phys. 59*, 5068 (1973).

F10.  J. H. Richardson, L. M. Stephenson, and J. I. Brauman, Photodissociation of $Fe(CO)_4^-$ and $Fe(CO)_3^-$ in the gas phase, *J. Amer. Chem. Soc. 96*, 3671 (1974).

F11.  K. J. Reed and J. I. Brauman, Photodetachment of electrons from Group IVa binary hydride anions: the electron affinities of the $SiH_3$ and $GeH_3$ radicals, *J. Chem. Phys. 61*, 4830 (1974).

F12.  M. S. Foster and J. L. Beauchamp, Chemical consequences of electron scavenging by $SF_6$ in radiolysis experiments, *Chem. Phys. Lett. 31*, 479 (1975).

F13.  J. F. G. Faigle, P. C. Isolani, and J. M. Riveros, The gas phase reaction of $F^-$ and $OH^-$ with alkyl formates, *J. Amer. Chem. Soc. 98*, 2049 (1976).

# Author Index to the Bibliography

Marshall, A. G.,   A36, A37,
  A38, A39, C15, C26
Marx, R.,   C35, F8
Massey, R. I.,   C22
Mauclaire, G.,   C35, F8
Melby, E.,   E65
Miller, J. S.,   A43, B41
Miyamae, T.,   E34
Morton, T. H.,   E38
Mosesman, M.,   C16, C23

Nibbering, N. M. M.,   E61
Nixon, W. B.,   B12
Norbeck, J.,   C24

Odom, R. W.,   C68
Olah, G. A.,   E43, E44, E65
Olmstead, W. N.,   E67
O'Malley, R. M.,   A7, A10,
  A11, A15, B35, C19, C22,
  E13, E14, E24, E58, E59

Parker, C. E.,   E64
Patt, S. L.,   E37
Pearson, P. K.,   D15
Pesheck, C. V.,   E70
Pinizzotto, R. F., Jr.,   C61,
  E55
Ponnamperuma, C.,   E12
Pople, J. A.,   B44
Prinstein, R.,   B6

Reed, K. J.,   F11
Richardson, J. H.,   D15, F9,
  F10
Ridge, D. P.,   A5, B26, E31,
  E68, E69
Riggin, M.,   A32, A44, A45,
  A46, C70
Riveros, J. M.,   B14, B28,
  B29, B32, B34, B36, B37,
  E54, E66, F13

Sammons, M. C.,   E64
Schaefer, J.,   C8, E8
Schleyer, P. v. R.,   B44
Scott, J. A.,   B32

Sefcik, M. D.,   C65, C67
Shaefer, H. F., III,   D15
Sharp, T. E.,   C34
Shen, J.,   E43, E44, E65
Silvers, J. H.,   B38, E63
Simms, W. T.,   A31
Smith, D. L.,   B46, C36, C43,
  C68
Smyth, K. C.,   A18, B17, F1,
  F2, F3, F4, F5
Sohma, J.,   A34, E34
Stahley, R. H.,   B42
Stephenson, L. M.,   D15, F9,
  F10
Stewart, G. W.,   C59, E47
Strong, J.,   B8
Su, T.,   B45, C41, C42, C51,
  C52, C53, C55
Summerhays, K. D.,   B33

Taagepera, M.,   B22, B23, B25,
  B33
Taft, R. W.,   B19, B22, B23,
  B25, B33, B49
Tannenbaum, H. P.,   E42
Tesarek, J. M.,
  see Bursey, J. M.
Tiedemann, P. W.,   B28, B36,
  B37, E54, E66
Tomer, K. B.,   D13
Tripodi, M. K.,   C67, E47

van der Hart, W. J.,   C63

Wallace, R. W.,   A18, F2
Wanczek, K. P.,   A33
Webb, H. M.,   B15, B24, B31,
  B47, B50
Wexler, S.,   C10
White, M. J.,   B17, C50
Wien, R. W.,   C29
Wilkins, C. L.,   A21, A22, D14
Wolf, J. F.,   B49
Wolfsberg, M.,   B49
Woodgate, S.,   A12, E18, E37
Woods, I. B.,   A32, A46

# APPENDIX 1 Calculation of Rate Constants for Experiments in the Conventional Drift Cell[1]

In mass spectrometers built or modified for the study of ion-molecule reactions, the formation, reaction, and analysis of the ions occur in different parts of the instrument (they are space-separated), as suggested by Figure 1-2. In an ICR spectrometer the primary ions are formed at the electron beam, but the ions resulting from chemical reactions are formed throughout the source and analyzer. Thus ions are formed, detected, and consumed in the analyzer. With so much happening simultaneously, the calculation of rate constants is much more complicated than in the case of mass-spectrometric studies. It can only be done indirectly by deriving an equation for signal intensities in terms of the rate constants, and then adjusting approximate rate constants until the calculated and observed ion intensities agree.

The raw data for the calculation of rate constants in experimentation on an unmodified instrument are the signal intensities of reactant and product ions, the magnetic field strength at which a certain ion is in resonance, and the static electric field intensities in the source and analyzer. The number densities of the neutral molecules are also known. The exact times required by the ions to drift through the source and analyzer regions of the cell must be known. As they depend on the uniformity of the static electrical fields in the cell, a small instrumental modification is made to ensure maximum uniformity of the field in the analyzer. The ion collector is disconnected from the electrometer and its four plates are wired so that each carries the same potential as the analyzer plate next to it. Thus the static field within the analyzer is extended through the ion collector.

The familiar rate equation for a bimolecular reaction, in

which the formation and destruction of the ion or molecule of interest is expressed by terms of the form $\pm k(A)(B)$, is here more conveniently written as $\pm kNF_r$, where N is the number density of the neutral molecule that participates in that bimolecular reaction and $F_r$ is the ion current. It is assumed that all reactions are bimolecular.

The time t required for an ion to drift from the electron beam to point x in the analyzer is given by

$$t = \frac{x_s}{v_s} + \frac{x}{v_a} \qquad (1)$$

where $v_s$ and $v_a$ are the ion-drift velocities in the source and analyzer respectively. This simple relationship between the time and the position of an ion permits integrations over the length of the analyzer to be written instead as integrations over time of passage through the analyzer. Thus the equation for the signal intensity of the ion i, which is given by

$$I_i = \int A_i(t) \; F_i(t) \; dt \qquad \text{(Equation 12, Chapter 1)}$$

will contain only functions of the time, even though it will be helpful to think of summing the contributions to an ion current across the analyzer. The limits of integration will depend on instrumental parameters.

We show in Table 1 the various symbols used in the derivation of the equation for signal intensity, and in Figure A-1 the formalism for describing what happens in different regions of the cell.

The ion current $F_r(t)$ of any ion whatsoever will vary with time according to the equation

$$\frac{dF_r(t)}{dt} = \sum_n \sum_q k_{nqr} N_n F_q(t) - \sum_m \sum_s k_{mrs} N_m F_r(t) \qquad (2)$$

where the first double sum is over all the bimolecular reactions between neutral molecules of species n and ions of species q and order one less than $F_r$ that react to form the ion $F_r$. The second double sum is over all the reactions of neutral molecules of species m with $F_r$ to form the ionic products of species s; as the terms in this double sum consume ions of species $F_r$, its sign is negative.

Equation (2) is an aid to computation, representing what would happen if a certain number of primary ions were formed

Table 1.  Symbols Used in This Appendix

---

$A_i(t)$  Instantaneous power absorbed by the ion i

$\varepsilon_1$  Amplitude of the radiofrequency field with which the
ion is in resonance

$F_r$  Ion current of the ion r, formed by a series of
(F - 1) reactions starting with a primary ion

f  Counting index for ion orders (primary, secondary,
tertiary)

$I_i$  Signal intensity of the ion i

i  Counting index for an ion without reference to a
reaction scheme

$K_i$  Constant defined by A(t)/t in Equation (7), page 16

k  Bimolecular rate constant; its units are typically
$cm^3$ molecule$^{-1}$ $s^{-1}$

$k_{nqr}$  Rate constant for the reaction of the neutral
molecules n with the ions q to form the ions r

$m_i$  Mass of the ion i

$N_m$  Number density of the molecule m

$P(t)$  Current of a primary ion

r  Counting index for ions of order f

t  Time

t'  A time prior to t; the ions formed in the analyzer
during t' + dt' react during t + dt (see Figure
A-1)

$\tau$  Drift time through the source

$\tau'$  Drift time through the analyzer

$v_a$  Drift velocity of ions in the analyzer

$v_s$  Drift velocity of ions in the source

x  Position coordinate in the analyzer, taken parallel
to the length of the cell (see Figure A-1)

$x_s$  Length of the source region

---

during a brief interval of time and if these drifted through the
cell and reacted and were detected before more ions were formed
at the electron beam.  We shall use this mathematical device
before we come to Equations (5) and (8), where integration over
the time the ions reside in the cell will bring the equations
back to instrumental reality.  A steady state exists in the
drift cell, in which nothing *measurable* varies with time.

Primary ions are, by definition, formed only at the electron beam, and not by any chemical reactions.  For these ions,
the first double sum in Equation (2) is zero, and the equation
becomes

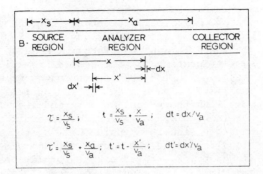

Figure A-1. Schematic diagram of the cell and the variables of ion motion in it. The cell is drawn to scale. B indicates the electron beam that forms the primary ions.

$$\frac{dP(t)}{dt} = -\sum_{m} \sum_{s} k_{mps} N_m P(t)$$

The ionic index on the rate constant is now p. The rate constants and molecular number densities are independent of time, so the integration is simply

$$\int_0^t \frac{1}{P(t)} \frac{dP(t)}{dt} dt = -\sum_{m} \sum_{s} k_{mps} N_m \int_0^t dt$$

which gives for the ion current of a given primary ion as a function of time,

$$P(t) = P(0) \exp(-\sum_{m} \sum_{s} k_{mps} N_m t) \tag{3}$$

Using Equation (1) and its context, we can write

$$P(t) dt = P(0) \exp(-\sum_{m} \sum_{s} k_{mps} N_m t) dt \tag{4}$$

for the number of primary ions that exist at time t in the

region x, x + dx of the cell.  Because we are dealing with a
*current*, it is also the expression for the number of primary
ions that would pass a stationary counter at some point x of the
analyzer between times t and t + dt.

Expressions for the primary, secondary, tertiary, and
higher ion currents can be written for any specified reaction
scheme.  They are the $F_r(t)$ values of Equation (2).  Evaluation
of the double sum in Equation (3) for a given reaction scheme
would produce the equation for the corresponding primary ion
current.  Let the $F_r(t)$ be such a set of ion current equations.

Because ions, with the exception of primaries, can be
formed in both the source and analyzer, but are detected only in
the latter region, the signal intensity will be the sum of two
terms, one for ions formed in the source and the other for those
formed in the analyzer.  $F_r(\tau_{fr})dt$ source-formed ions enter the
analyzer during a time dt.  The number of them that remain when
they reach the region x, x + dx is

$$F_r(\tau_{fr}) \, \exp\left( \sum_m \sum_s k_{mrs}N_m(t - \tau_{fr}) \right) dt$$

They have been in the analyzer for a time $t - \tau_{fr}$; thus the
instantaneous power absorbed per ion is $K_{fr}(t - \tau_{fr})$, where K is
given in Table 1.  By Equation (12) of Chapter 1, these expressions for ion current and power as functions of time lead to an
equation for the signal intensity of ions formed in the source,
of

$$I_{fr}(\text{source-formed}) =$$

$$K_{fr}F_r(\tau_{fr}) \int_{\tau_{fr}}^{\tau'_{fr}} (t - \tau_{fr}) \, \exp\left(-\sum_m \sum_s k_{mrs}N_m(t - \tau_{fr})\right) dt \qquad (5)$$

To calculate the contribution by ions formed in the
analyzer to the signal intensity, it is first necessary to
characterize their rate of formation in a differential fashion.
Let t' be some time prior to the arrival of these ions in the
region x, x + dx of the analyzer.  The number of ions formed
there between previous times t' and t' + dt' is, according to
the double sum for ion formation in Equation (2),

$$\sum_n \sum_q k_{nqr}N_nF_q(t') \, dt'$$

The number of these ions that reach the region x, x + dx before undergoing further reaction is

$$\sum_n \sum_q k_{nqr} N_n F_q(t') \, dt' \, \exp\left(-\sum_m \sum_s k_{mrs} N_m (t - t')\right) \qquad (6)$$

These ions will have been in the analyzer for time $(t - t')$, so their power per ion is

$$K_{fr}(t - t') \qquad (7)$$

The contribution to the signal intensity from ions formed during $dt'$ and drifting through $x$, $x + dx$ is the product of Equations (6) and (7). Integration over previous formation times in the analyzer gives

$$I_{fr}(x, \, x + dx, \, \text{analyzer-formed}) =$$

$$K_{fr} \int_{\tau_{fr}}^{t} (t - t') \sum_n \sum_q k_{nqr} N_n F_q(t') \, \exp\left(-\sum_m \sum_s k_{mrs} N_m (t - t')\right) dt'$$

The total signal intensity due to the ions in all regions of the cell is obtained by integrating between the analyzer entrance and exit times of the ion of interest:

$$I_{fr}(\text{analyzer-formed}) =$$

$$K_{fr} \int_{\tau_{fr}}^{\tau'_{fr}} dt \int_{\tau_{fr}}^{t} (t - t') \left( \sum_n \sum_q k_{nqr} N_n F_q(t') \right) \times$$

$$\exp\left(-\sum_m \sum_s k_{mrs} N_m (t - t')\right) dt' \qquad (8)$$

The detection system does not distinguish between ions formed in the source and those formed in the analyzer, so the intensity of the $r^{th}$ ion of order F is the sum of Equations (5) and (8):

$I_{fr} =$

$$K_{fr}F_r(\tau_{fr}) \int_{\tau_{fr}}^{\tau'_{fr}} (t - \tau_{fr}) \exp\left(-\Sigma \Sigma_{m\ s} k_{mrs}N_m(t - \tau_{fr})\right) dt +$$

$$K_{fr} \int_{\tau_{fr}}^{\tau'_{fr}} dt \int_{\tau_{fr}}^{t} (t - t')\left(\Sigma \Sigma_{n\ q} k_{nqr}N_n F_q(t')\right) \times$$

$$\exp\left(-\Sigma \Sigma_{m\ s} k_{mrs}N_m(t - t')\right) dt' \tag{9}$$

The integrals in Equation (9) can be evaluated for a particular set of reactions by inserting the correct ion currents. The result is an algebraic expression for the signal intensity of a particular ion in terms of all of the rate constants and molecular number densities of the reactions that form and consume it, and its drift times through the cell.

Except in special cases, these signal intensity equations cannot be solved for rate constants, hence rate constants are calculated from an approximate equation and substituted into the algebraic equation for the signal intensity. The intensity thus calculated is compared with the observed intensity, and an iterative program is used to compute refined rate constants.

This is the most general method of extracting reaction-rate constants from ICR data in the conventional drift cell.

In order to apply Equation (9) it is necessary to know the drift times of the ions through the source and analyzer. Such times can be calculated from Equation (1). An indirect measurement of the drift times in the modified cell described earlier (p. 210) gave satisfactory results, but in unmodified cells the true drift times, which depend on the drift potentials, are not necessarily obtained in so simple a fashion.[A20,C22]

NOTE

1. Except for the preliminary remarks, this appendix is taken from the work by Marshall et al.[C15] It has not been revised in the light of more recent work, nor has the notation been brought into exact agreement with the rest of the book.

# APPENDIX 2  On the Theory of the
# Marginal Oscillator

The following excerpt is from Robert T. McIver, Jr., A solid-
state marginal oscillator for pulsed ion cyclotron resonance
spectroscopy, *Rev. Sci. Instrum. 44*, 1071 (1973). Equation (8)
shows that the power absorption A (and not the energy or the ion
current, for example) determines the signal height $\Delta V$. The
literature references are reproduced below.

6.  D. Wobschall, J. Graham, Jr., and D. Malone, *Phys. Rev.*
    *131*, 1565 (1963).
10. R. A. Brooks, *Rev. Sci. Instrum. 43*, 807 (1972).
24. M. S. Alder, S. D. Senturia, and C. R. Hewes, *Rev. Sci.*
    *Instrum. 42*, 704 (1971).
25. M. S. Alder and S. D. Senturia, *Rev. Sci. Instrum. 40*, 1481
    (1969).
26. M. B. Comisarow, *J. Chem. Phys. 55*, 205 (1971).
27. W. T. Huntress, Jr., *J. Chem. Phys. 55*, 2146 (1971).
28. S. E. Buttrill, Jr., *J. Chem. Phys. 50*, 4125 (1969).
29. J. L. Beauchamp, *J. Chem. Phys. 46*, 1231 (1967).

## THEORY

"The sensitivity of marginal oscillator-type circuits has been
thoroughly discussed in recent papers.[10,24,25] This section
briefly reviews the pertinent considerations for achieving
maximum performance and includes some particular details applicable to the pulsed marginal oscillator circuit.

"Figure 1(a) shows the circuit diagram of a basic feedback
oscillator. The parallel GCL circuit has a resonant frequency

$$\omega_0 = (LC)^{-1/2}, \tag{2}$$

where L is the inductance and C the capacitance of the resonant
circuit. The conductance G represents power losses due to the

nonideality of the circuit. The limiter provides a constant amplitude rf voltage output which is coupled to the resonant circuit through the feedback resistor, $R_f$. At the resonant frequency, $\omega_0$, the admittance of the resonant circuit is a pure conductance and

$$V_1 = \frac{V_0}{1 + GR_f}, \tag{3}$$

where $V_0$ is the rms voltage output of the limiter, and $V_1$ is the rms voltage across the resonant circuit. If the load is increased by an incremental amount, $\Delta G$, the voltage across the resonant circuit becomes

$$V_1' = \frac{V_0}{1 + (G + \Delta G)R_f}. \tag{4}$$

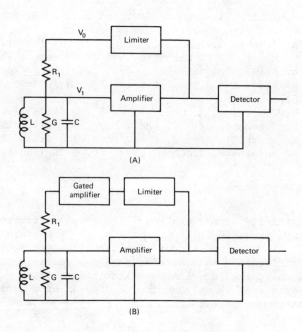

Fig. 1.  Basic marginal oscillation circuits.
(a) Limited self-oscillator, and (b) pulsed marginal oscillator.

The incremental conductance $\Delta G$ represents the icr absorption.
Taking the 'signal' to be $\Delta V = V_1 - V_1'$ gives

$$\Delta V = V_0 \left( \frac{1}{1 + GR_f} - \frac{1}{1 + (G + \Delta G)R_f} \right) \tag{5}$$

$$= V_1 \Delta G \left( \frac{R_f}{1 + (G + \Delta G)R_f} \right) . \tag{6}$$

For small signals $\Delta G$ is quite small, and Eq. (6) can be approximated by the form

$$\Delta V = V_1 \Delta G \left( \frac{R_f}{1 + GR_f} \right) . \tag{7}$$

The incremental conductance, $\Delta G$, gives rise to a power absorption, $A = V_1^2 \Delta G$, for $V_1 \simeq V_1'$. Equation (7) can be rewritten in terms of power absorption to give

$$\Delta V = \frac{A}{V_1} \left( \frac{R_f}{1 + GR_f} \right) . \tag{8}$$

"The power absorbed by a gaseous ion in the presence of a homogeneous magnetic field and an rf electric field can be calculated.[6,26-29] Comisarow[26] has derived the result

$$A = \frac{q^2 E^2}{4m\xi} \{1 - \exp(-\xi t)\}, \tag{9}$$

where A is the instantaneous power absorption at resonance $\omega_c = \omega_0$), E the rf electric field, $\xi$ the reduced collision frequency for the ion, and t the time the ion has been exposed to the rf electric field.
"Combining Eqs. (8) and (9) shows that the signals observed in ion cyclotron resonance experiments are (1) linearly proportional to the number of ions of a particular charge-to-mass ratio and to the oscillation level, $V_1$, of the marginal oscillator, and (2) inversely proportional to the mass of the ion detected."

# Index

221